FOUR TREATISES
OF THEOPHRASTUS VON HOHENHEIM
CALLED PARACELSUS

THEOPHRASTUS VON HOHENHEIM
CALLED PARACELSUS

Woodcut of 1538, by Augustin Hirschvogel

FOUR TREATISES

OF

THEOPHRASTUS VON HOHENHEIM
CALLED PARACELSUS

Translated from the original German, with Introductory Essays

BY

C. LILIAN TEMKIN · GEORGE ROSEN
GREGORY ZILBOORG · HENRY E. SIGERIST

EDITED, WITH A PREFACE
BY
HENRY E. SIGERIST

BALTIMORE AND LONDON
THE JOHNS HOPKINS UNIVERSITY PRESS

Originally published in a hardcover edition by the Johns Hopkins Press, 1941
Johns Hopkins Paperbacks edition, 1996
05 04 03 02 01 00 99 98 97 96   5 4 3 2 1

The Johns Hopkins University Press
2715 North Charles Street
Baltimore, Maryland 21218-4319
The Johns Hopkins Press Ltd., London

Library of Congress Cataloging-in-Publication Data

Paracelsus, 1493–1541.
  Four treatises of Theophrastus von Hohenhein, called Paracelsus / translated from the
original German, with introductory essays by C. Lilian Temkin . . . [et al.] ; edited,
with a preface by Henry E. Sigerist.—Johns Hopkins paperbacks ed.
    p.   cm.
  "Originally published in a hardcover edition by the Johns Hopkins Press, 1941"—
Copr. p.
  Includes bibliographical references (p.   ).
  ISBN 0–8018–5523–3 (pbk. : alk. paper)
  1. Medicine—Early works to 1800.   2. Medicine, Magic, mystic, and spagiric—
Early works to 1800.   3. Occultism—Early works to 1800.
I. Temkin, C. Lilian (Clarice Lilian), 1906–   .   II. Sigerist, Henry E. (Henry Ernest),
1891–1957.   III. Title.
R128.6.P2213   1996
610—dc20   96-30930
                                                                    CIP

A catalog record for this book is available from the British Library.

TABLE OF CONTENTS

PREFACE

Four hundred years ago, on September 24, 1541, Paracelsus died in Salzburg. He died misunderstood by the world, embittered, and poor. With him went one of the most forceful personalities of the Renaissance.

Born near Einsiedeln toward the end of the year 1493, Philip Theophrastus Bombast von Hohenheim, who later called himself Paracelsus, was the son of a physician. His father Wilhelm von Hohenheim was the impoverished scion of a noble family of Suabia. He had married a Swiss girl and practised medicine on the pilgrims' road that leads to the Benedictine Abbey of Einsiedeln. Thus Paracelsus grew up in the mountains of Switzerland until the family, in 1502, moved to Carinthia where the father had accepted a position as municipal physician of the mining town of Villach.

Paracelsus wanted to be a physician like his father. He studied the arts, possibly in Vienna, then medicine in Italy at the University of Ferrara. But he was thoroughly disappointed with the instruction that was given at the time. From his father he had learned to see nature with his own eyes, not through the medium of books, and he had learned much biology with him. In Villach, laboring in mines and smelting works, he had acquired a thorough knowledge of chemistry. And now, in Italy, he found that there was too wide a gap between science and the Graeco-Arabic theories that were still the foundation of medicine. He was a rebellious spirit, at that time already hard-headed and stubborn. Since he could not find at the university the enlightenment he was seeking, he looked for other teachers and started on a voyage of discovery which, with short interruptions, was to last until his death.

For years he travelled all over Europe including the British Isles, went as far north as Sweden, east to Poland and Lithuania, to Turkey, Greece, the Ionian islands and Alexandria. During his wanderings he practised medicine as an itinerant doctor, visited mines and workshops, studied local diseases, and was eager to learn from any source, humble as it might be.

In these years of wanderings, meditating over the problems of life and death, of health and disease, and of the world at large, he developed his own system of medicine and also a philosophy and theology of his own. He conceived disease in different terms than heretofore and applied in his therapy drugs, particularly chemical compounds, many of which had not been used before. A scientist and mystic, he tried to determine the place of man in the world, his relation to God, the causes of disease and its meaning. More and more he felt that he had a mission to fulfil and a message to carry to the people. Medicine had degenerated and was ripe for reformation. He was to be the Luther of medicine. And since the people refused to listen, he yelled his message at them and became more and more bitter and aggressive.

He wrote a great many books that cover a wide range of subjects, but there was a deliberate conspiracy against him and only a few of his works, and not the most important, were published during his lifetime. Whenever a publisher was willing to take a chance, organized medicine—in those days the medical faculties—brought pressure on him, threatening boycott, and the attempt was frustrated.

Once he was given an opportunity to teach students and thus to spread his doctrines by word of mouth. In 1527 he was offered the position of muncipal physician in Basle, which included the privilege of giving courses at the university. He accepted eagerly and gave courses from 1527 to 1528. He was soon disappointed in his high expectations. Antagonized by his colleagues, not understood by the students, he soon had quarrels with everybody and left the city suddenly.

He resumed his wanderings. Several times he attempted to settle down, always hoping to have his books printed, but circumstances always drove him away. When he came to Salzburg in 1541, he was 48 years of age but an old man, sick and worn.

His contemporaries knew little about him. He had few faithful disciples able to understand his thought and to endure his rough manners. He undoubtedly had the reputation of an able physician who knew potent drugs and was particularly skillful in the treatment of wounds and other surgical diseases. Syphilis was one of them, and the only major works published during his life were the *Grosse Wundarzney* and a treatise on syphilis. The violent tirades he directed against physicians missed their target because they were not printed. His philosophical and theological writings could have been known to only a very few people. He did not become the reformer of medicine. Two years after his death, in 1543, the Fabrica of Vesalius appeared, a book that inaugurated a reformation of a different kind, slower but more effective. Like a comet, Paracelsus went through the lands. His erratic ways and strange habits gave rise to legends while he was still alive, and to many he appeared as a mysterious Doctor Faustus who could perform miracles.

A more correct appreciation began when some of his disciples, notably Adam von Bodenstein and Johannes Huser, published his manuscripts and manuscript copies of his works. In the second half of the 16th and throughout the 17th century it was particularly his chemistry and chemotherapy that attracted wide attention. Van Helmont was strongly influenced by him, and the Iatrochemists drew freely from his works. Many of his remedies became generally accepted. And then, as today still, alchemists, astrologers, religious cranks and other mystics worshipped him as a master into whose works they could read whatever pleased them. Whenever medicine went through a mystic speculative phase, as was the case in Germany around 1800, and again today, the figure of Paracelsus came to the fore. He is very German in his desperate struggling for a philosophy that would unriddle heaven and earth, and also in his unbalanced aggressiveness. There is no greater contrast than between him and a thinker like Descartes.

A study of Paracelsus is still very stimulating. He undoubtedly was one of the most original personalities in a period that had produced many great characters, and whoever wants to have a picture of the Renaissance cannot ignore him. To the physician his approach to the problems of health and disease is very interesting. In his endeavor to establish a philosophy of medicine he attacked many problems that are still unsolved today. He was a spiritualist, and if we are rationalists we will feel closer to Descartes and to the 18th century but we cannot deny that Paracelsus drew our attention to many questions that still need investigation.

The only access to Paracelsus' thought is through his works, but the study of his books is very difficult. They were written not in Latin but in German, and not in modern high German but in the 16th century German of Switzerland. It was a language in which it was still difficult to express scientific matters, and even those who master the language will find many obscure passages. Translations into modern languages, therefore, require no justification. They are more than mere versions from one language into another; they are interpretations.

In commemoration of the four hundredth anniversary of the death of Paracelsus we are presenting English translations of four treatises that illustrate four different aspects of his work.

The first, a translation of *Sieben Defensiones, Verantwortung über etliche Verunglimpfungen seiner Missgönner,* is probably Paracelsus' most personal work. It is a passionate justification of his character, activities and views, and gives a splendid picture of the man and of his basic ideas. We were particularly anxious to make this treatise available to the English-speaking public and C. Lilian Temkin's translation has succeeded in being accurate while still preserving the flavor of the original style.

The second treatise is a translation of *Von der Bergsucht und andern Bergkrankheiten drei Bücher,* a study on the diseases of miners. It is the first monograph ever written on the diseases of an occupational group and thus holds a very important place in the history of medical literature. Paracelsus had more practical experience in miners' diseases than any physician of the time and his observations are very interesting. The book became the starting point of a new line of medical literature. Dr. George Rosen, one of the most promising of the younger medical historians, who has written a splendid— unfortunately still unpublished—book on the history of the diseases of miners and who has a thorough knowledge of German, was best prepared to undertake the translation and interpretation of this treatise.

Then follows a translation of the treatise *Von den Krankheiten, so die Vernunfft berauben* by Dr. Gregory Zilboorg. It is the most important document on the psychology and psychiatry of Paracelsus. Written at a time when mental diseases were again to be studied and treated by physicians, it is a pioneering book also, and anticipates modern views. Dr. Zilboorg has just completed a book on the history of medical psychology and he previously wrote a study on *The Medical Man and the Witch During the Renaissance* [1] so that he is the foremost expert in the psychiatry of that period.

The last contribution, finally, is a translation of the *Liber de nymphis, sylphis, pygmaeis et salamandris et de caeteris spiritibus*. In spite of its Latin title the treatise was written in German. I became interested in it, years ago, in the course of studies on the survival of paganism in Christianity, and I was soon fascinated by its poetical character. All of us who had a classical education, whatever our present interests may be, have deep down in our hearts a nostalgia for the Greek world that was inhabited by such lovely creatures as the nymphs. I studied Paracelsus' treatise primarily from the point of view of comparative literature and art, but I found that it is also a good sample of his philosophy and theology. And so I decided to include it in this volume.

We all had to struggle with the difficulties presented by the Paracelsian language and style. A translation into modern English would destroy Paracelsian thought. And so we all tried to be as intelligible as possible while still following the original text as closely as we could. Each one of us has solved the problem in his own way.

---

[1] The Johns Hopkins Press, Baltimore, Md., 1935.

The four introductory essays preceding the translations were read at a Symposium on Paracelsus held at the Seventeenth Annual Meeting of the American Association of the History of Medicine in Atlantic City, on May 5, 1941.

The Literature at the end of the book intends to be a short guide to those readers who would like to learn more about Paracelsus.

In publishing this book we wish to contribute our share in reviving the personality of an honest man who was a great physician and a staunch fighter for what he considered the truth. It is so easy to be orthodox and to reap honors by repeating what people expect and wish to hear. Progress, however, is achieved through the clash of ideas, and heretics like Paracelsus are a ferment without which there would be no life.

HENRY E. SIGERIST

THE JOHNS HOPKINS INSTITUTE
OF THE HISTORY OF MEDICINE
September 24, 1941

I

SEVEN DEFENSIONES

THE REPLY TO CERTAIN CALUMNIATIONS OF HIS ENEMIES

BY

THEOPHRASTUS VON HOHENHEIM

CALLED PARACELSUS

TRANSLATED FROM THE GERMAN, WITH AN INTRODUCTION

BY

C. LILIAN TEMKIN

INTRODUCTION

In 1538, ten years after leaving Basel, Paracelsus settled down again for a few years in the Austrian province of Carinthia, making it his second home. The intervening years of wandering had seen clashes with the medical faculties of the universities of Leipzig and Vienna, in which they successfully prevented the printing of his works. Now, however, he feels he is in sympathetic surroundings and it is therefore not hollow praise which he pours on the province of Carinthia, when it promises publication of three works which are now known as the Carinthian writings :—

The First Book: The Reply to certain Calumniations of his Enemies, or, as it is more commonly called, The Seven Defensiones.

The Second: On the Errors and Labyrinth of the Physicians, that they should learn in other books than heretofore.

The Third: On the Origin and Cause of Sand and Stone, also the Cure of the Same.[1]

The result of his praise of his apparent patrons is pathetic indeed. There is an urgency pervading the writing, and yet, in spite of the promises to print it immediately, the manuscript lay idle for twenty-five years. It took the enthusiasm of a Paracelsist of the next generation, Dr. Theodor Byrckmann, backed by an official enquiry from the government at Vienna, to find the lost manuscript, which was then published in Cologne in the printing house of Dr. Byrckmann's father, Arnold Byrckmann, in 1564.[2]

The first of these Carinthian writings is now offered in English translation: the Seven Defensiones, which contain

[1] Cf. Huser, p. 143.
[2] Cf. Sudhoff (1928), Vorwort, pp. viii-xi.

3

in comparatively short space the quintessence of Paracelsian philosophy.

At the age of forty-five, Paracelsus feels himself prematurely aged. All his life he has been at odds with medical authorities and he feels he must once and for all answer their charges and defend his own way of thought. It is obvious that such a defence is more than likely to turn into an attack on his opponents and Paracelsus gives full vent to his anger against the traditionalists, the scholastic physicians, the " sophistic legion," as he calls them, and even the humanists whose quibbling concerning terminology arouses only his scorn. He conceives of the noble heritage of medicine as stemming from Apollo, Machaon, Podalirius and Hippocrates, who worked wonders in the Light of Nature. This Light of Nature is the light which is in nature, revealing her inmost being and hidden depths; it is hence the guiding principle of all knowledge of nature.[3] But medicine has fallen among the anti-physicians and its delivery out of their hands is the mission which Paracelsus takes upon himself. And indeed, in the light of this mission, the crimes of which his opponents accuse him must inevitably appear as essential virtues.

He has been accused of lack of respect for medical tradition and he ridicules respect for a medical system because of its age alone, for new times require a new medicine. This determination of Paracelsus to work in the present and disregard the past is reflected in his astronomical beliefs. For the medieval astrologer, as well as for Paracelsus, the firmament is not only a measure of time, but inextricably associated with all events in time and, indeed, it varies its influence as it grows older. Paracelsus believes that each age is endowed with its own potentialities and its own destiny, and that the sole duty of man is to work within his own age to realize its potentialities and fulfil its destiny. At the same time he apparently uses the current historical terminology of the ' world-monarchies,' a con-

[3] Cf. Bodo Sartorius Freiherr von Waltershausen, *Paracelsus am Eingang der deutschen Bildungsgeschichte*, Leipzig, 1936, p. 104.

cept derived from the division of human history into four 'monarchies' by St. Jerome.[4] He abandons the implied idea of limiting world history to but four periods of which the last was synonymous with the Roman Empire—a limitation long obsolete at his time but retained by the majority of historians—and speaks of 'many' monarchies, again imbuing an old term with the new meaning of 'epoch.'

Not by their age but by their fruits, i. e. their works, are medical systems to be judged, according to Paracelsus. The false physician he sees hiding his ignorance behind senseless activities, or exploiting the credulity of the peasants to live on the fat of the land. The true physician, on the other hand, knows what to do and does it with expedition. He is honest in his dealings with his patients and earnest in his search for further knowledge. To him there are no incurable diseases, and indeed any such contention would seem impious to him. He knows God to be capable of curing all diseases, even the most striking and wondrous ones. To Paracelsus the practice of medicine is a supremely religious function. "The physician," he says, "is he who in the bodily diseases takes the place of God and administers for Him." (p. 15) And indeed the religious aspect of his philosophy colours not only his medical thought but also his literary expression. For him the physician must be motivated by the Christian injunction to love one's neighbour; he must acknowledge God as the only true fountainhead of the medical art and must accept God's disposal of natural resources in humility and strive to extend his knowledge of it. The truth is to be sought both in nature and in Holy Writ, and the Seven Defensiones themselves evince the constant interplay of the two sources of knowledge in Paracelsus' own medical system. Illustrations of medical axioms are inevitably drawn from the parables of Christ and this technique sometimes leads to a parallelism of argument which appears obscure to the modern reader, but which sheds considerable light on Paracelsian thought once its logic is fathomed.

[4] Cf. Ernst Bernheim, *Lehrbuch der historischen Methode und der Geschichtsphilosophie*, Leipzig, 1903, pp. 66-67.

Paracelsus defends his introduction of new names and new prescriptions and his description of new diseases. The neglect of these new diseases by older physicians he ascribes categorically to their ignorance of astronomy. The new names he justifies on the ground that the old are too complicated as to linguistic derivation—not that he does not understand each component part, for any simple scholar can read such names as are given by the ancients—but the composite terminology is meaningless and is, moreover, distorted from village to village. Paracelsus prefers to rename diseases, seeking not rhetoric or Latin in them, but medicine of which the old names gave him no account.

But the greatest outcry against Paracelsus has been directed against his new prescriptions. He has been accused all his life of using poison and this accusation is aimed at the chemical aspect of his medical system. Paracelsus was adamant in his contention that the physician must be a chemist, and the opposition to his practices long outlived Paracelsus himself. The use of inorganic, particularly metallic, elements in internal remedies was attacked as unnatural and poisonous for more than a century and the argument found one of its most acute expressions in the famous fight of the Paris faculty about antimony. Paracelsus refutes the contention on the ground that all things are poisonous if taken without respect to their proper dosage. Further, he accuses his opponents of using poisons too, and, what is worse, of not knowing the proper dosage.

Paracelsus is accused of being a wayfarer, of being resident nowhere; but of this fact he is proud rather than ashamed. Indeed, wayfaring is for him not only a way of life but the expression of a philosophy. It is the only valid way to increase the physician's knowledge; moreover, this knowledge, ' Erfahrung,' is to be understood not only in the secondary sense of ' experience,' but in the primary sense too of ' experience gained by travel.' [5] Only by the evidence of his senses can he be truly instructed; he must seek out nature and enquire into new dis-

[5] Cf. Friedrich Gundolf, *Paracelsus,* Berlin, 1927, pp. 16-20.

eases, for his teacher is not to be found in his own chimney-corner. The serenity of Paracelsus' convictions in this respect is reflected in his memorable words on the Codex Naturae at the end of the Fourth Defence.

The false physicians are pictured always as growing rich; they are spoilt by the ignorance of the people and do all for selfish gain, nothing for love, as the commandment bids. In this exaggerated condemnation Paracelsus seems to know of no exceptions amongst his predecessors and contemporaries since Hippocrates; but his venom singles out the Jewish physicians for an attack which voices the medieval prejudice against their people, just as does Shakespeare's portrayal of Shylock. As he scorns the motives of the " false " physicians, so he scorns their manners. They accuse him of being a strange, surly fellow, but he is proud of the distinction and claims that civility is the least requirement of the physician. They cloak their lack of knowledge with worldly manners; he lays full stress on his works and demands that medicine be broad enough for the native manners of all practitioners. In this insistance on the part of Paracelsus on the performance of works as the only truly important function of the physician and on the works themselves as the only true criterion by which to judge his merit, we see the emergence of the Protestant ethic, in spite of Paracelsus' own Catholicism. It lends a deeper meaning to his reputation as 'Lutherus medicorum,' and the vigour of his language in spite of the limitations of the German of his day contributes further to the similarity between the two great contemporaries.

The apothecaries complain that he is strange and eccentric, yet all he does is demand good drugs and reject their Quid pro Quo, i. e. their convenient substitutes for drugs prescribed, and write short prescriptions which do not empty the apothecaries' boxes quickly enough to fill their pockets with gold at the accustomed rate. He is accused, too, of being slow at diagnosis, yet this he deems a virtue. Only the followers of the Galenic doctrine of the four humours diagnose hidden diseases rapidly, and with what dire results!

Thus does Paracelsus justify his ways " for the last time until further provocation, which, if God will, will also be returned blow for blow." He defends himself with deep sincerity, with a conviction founded on an almost naive combination of humility before God and nature, and pride in his own achievements. He demands of medicine and of the physician a program of Christian love and compassion, and a knowledge not to be culled from the scholastic authorities he despises, but from investigation and experimentation.

The Seven Defensiones are not without their difficulties for the translator, for the bitterness of the author is often condensed in the most rugged of metaphors and his style is at once abrupt and obscure, but the passion of this scientific revolutionary cannot but fire the imagination of the reader. At the same time, it must be admitted that some few passages defy lucid explanation and here I felt that too free and logical an interpretation would only falsify the whole picture. Early sixteenth century German was a very imperfect instrument of literary expression; it abounds in awkward turns of phrase, grammatically incomplete sentences, and utter inconsistency as to genders, the use of weak endings, etc. As such it presents the translator with a task of interpretation far exceeding the interpretive element implicit in the translation of a modern text. This element has been allowed such free rein in, for instance, the Latin version of the Seven Defensiones, that the two texts are scarcely comparable. I have resorted to this extreme of interpretation only in the few instances where an idiomatic expression or play upon words could not be rendered in any literal manner; nevertheless my translation lays no claim to exemption from the general character of ' interpretation ' which would seem inevitable in any approach to an understanding of Paracelsus, whatever the language used. I have attempted, however, to retain as much of the style and linguistic quality as possible, and to this end I have kept as many of the frequently interspersed Latin expressions as fitted easily into the English sentences. Where alterations in the Latin form would have been required, the Latin has been translated and the

English rendering capitalised, and all proper names have been given in English.

The translation has been based on the edition by Johannes Huser:

Der Buecher und Schrifften des Edlen . . . *Philippi Theophrasti Bombast von Hohenheim Paracelsi genannt, Ander Theil,* Basel 1589, pp. 157-190.

In addition I have consulted:

Byrckmann's edition: *Drey Buecher Durch den Hochgelerten Herrn Theophrastum von Hohenheim Paracelsum genant,* etc., Coeln 1564, pp. 1-52. (Facsimile edition, Leipzig 1928).

Aur. Philip. Theoph. Paracelsi . . . *Opera omnia medico-chemico-chirurgica,* Volumen primum, Genevae, 1658, pp. 248-264.

Theophrast von Hohenheim (Paracelsus), Sieben Defensiones, etc. Eingeleitet und herausgegeben von Karl Sudhoff, Leipzig 1915. (Klassiker der Medizin, herausgegeben von Karl Sudhoff, Bd. 24). This will be quoted as Sudhoff (1915).

Theophrast von Hohenheim gen. Paracelsus, Saemtliche Werke. I. Abteilung herausgegeben von Karl Sudhoff, 11. Band, Muenchen und Berlin 1928, pp. 125-160. This will be quoted as Sudhoff (1928).

THE REPLY TO CERTAIN CALUMNIATIONS OF
HIS ENEMIES

(SEVEN DEFENSIONES)

PREFACE TO THE READER BY THE MOST LEARNED GENTLEMAN
AUREOLUS THEOPHRASTUS VON HOHENHEIM,
DOCTOR OF BOTH MEDICINES

Reader, pay heed, that I may tell thee why these *Defensiones* have been written by me. Whereas God suffered the spirit of medicine to emerge in its fundamentals through Apollo, through Machaon, Podalirius [6] and Hippocrates, and suffered the light of nature to work without a darkened spirit, exceeding wonderful great works, great *Magnalia*,[7] great *Miracula*, were performed through the Mysteries, Elixirs, Arcana and Essences of nature, and medicine was marvellously conceived in a few pious men, as was told above. Whereas, however, the Evil One with his corncockles [8] and his weeds suffers nothing to grow for us in an undefiled wheatfield, medicine has been darkened by the first spirit of nature and has fallen among the anti-physicians and has become so entangled with persons and sophistries, that no one has been able to advance as far in the work as Machaon and Hippocrates did. And in medicine what is not proved with works has lost its disputation and gains still less by argument. Now, my Reader, pay heed: if an effective doctrine set itself against the sophistic legion, would it not be justified, in order that the work might floor the gossip? Guess then, Reader, of whom I speak. It is of the holy ones who give no signs. The throngs and the concourse might frighten many a one, so that he would refrain from shutting

----

[6] Machaon and Podalirius are the legendary physicians who fought in the Greek army before Troy.

[7] I. e. mighty works.

[8] Cf. Sudhoff (1915), p. 13, footnote.

the mouth of the chatterer. But the result and the retraced
steps prove that no store should be set by the concourse. From
it arises the error that Hippocrates must be a gossip and the
spirit of truth in medicine must become a sophistic chatterer.
For what is there that is too much for a gossip? Of this rabble,
some have let their mouths be overhasty and have defended
themselves in this with infamous words, and since they have
taken medicine into their mouths, they must defend themselves
with their mouths, which can do nought but vilify and slander.
Such *Lingua dolosa*[9] has had a tilt at me too. But it is necessary not to leave them unanswered in these things as they are
not built on the first rocks of medicine but have placed themselves on a kitchen rock,[10] and have forgotten the truth of the
medical art, and have borne me and others around in false
faces with their sophistic Fables. If a man were dedicated to
the first *Centrum*,[11] such abuse would not proceed from him.
Their best art is their rhetoric and its ilk, and the virtue apper-
taining to Pseudo-physicians. Wherefore, Reader, the replies
follow below, so that and in order that you may be able to
decide in this matter. Although it would not be necessary to
refute the contentions of such people: one might allow them
to remain poetic physicians, rhetorical writers of prescriptions
and nebulous preparers: in time people would tire of them.
In order, however, that it may be understood that a physician
is of no account without works and that the work of physicians
is not empty talk, for the sake of teaching these things have I
undertaken this. And so, dear Reader, hindered as I have been
from publishing my writings, I have presented this to Carinthia, the archduchy, so it would reach thee through these same
worthy gentlemen, wherever in the world thou mayest receive
it. For without this province, Reader, it would not come into
thy hands. Love, therefore, the *Theoricam* in this work, but still
much more the works of the art. Set down in Sanct Veit in

---

[9] I. e. deceitful talk.
[10] Probably means their medicine is based only on the apothecary's kitchen.
[11] For the meaning cf. Von Waltershausen, l. c., p. 58 ff.

Carinthia, on the 19th day of August of the lesser reckoning, 38.[12]

THE FIRST DEFENCE IN THE DISCOVERY OF THE NEW MEDICINE OF DOCTOR THEOPHRASTUS

That I here in this work introduce a new *Theoricam,* also *Physicam,* together with new Concepts which heretofore have never been held, nor understood by Philosophers, Astronomers, nor Physicians, comes to pass for the reasons which I shall now relate to you:— One reason, for instance, which is sufficiently proven, viz: that the old *Theorici* described the *Rationes* and *Causas Morborum* wrongly and inaccurately and thus introduced such error, and so confirmed this error, that it was held and considered just and incontrovertible. And it is so deeply rooted and so tended and preserved, that no one may any more seek an alternative, or the same is deemed an error. This I wish to make known to you, for I am bound to judge such a thing as great foolishness: since heaven is forever bearing and creating *Ingenia,* new *Inventiones,* new *Artes,* new *Aegritudines* in the light of nature, should not these then also be valid? Of what avail is the rain that fell a thousand years ago? That which falls at present avails. Of what avail to the present year is the sun's course of a thousand years ago? Does not Christ interpret how we are to judge this, saying: it is sufficient that the day bear its own yoke? That is as much as to say: it is enough that thou doest what the day brings and that thou dost further conclude that the morrow too will bear its own cares. Now since care walks alone and every day has xii hours, and every hour its separate action, what harm then does the twelfth hour do to the first, or of what disadvantage to the first is the twelfth? Thus everything is set down in its own *Monarchiam*[13] according to its time. And concerning the present should we trouble ourselves, not concerning the past. And each *Monarchia* is provided with the full light of nature. These are

---

[12] "Der mindern zal 38 [1538]," Sudhoff (1928), p. 126. The "numerus minor" indicates the number of years after 1500.

[13] Cf. Introduction, p. 4 f.

the miracles of God: to vary the light of nature in many *Monarchias* between the beginning and the end of the world, which has often been overlooked and men have not acted according to what is contained in these monarchies. Therefore I, by virtue of the present light of nature, and according to the predestined order of the present monarchy, I will be unrebuked by anyone in my writing, and still less attacked and hindered on account of sophistry which I call an error in medicine.

Their foolishness I must the better expose for the sake of making known my reasons and their error, and in this the universities will not overthrow me and this, then, I make known to them. Medicine is a work and since it is a work, the work will testify to its master. Now behold in the works how every part is known and judged: the work is an art, the art gives the doctrine of the work, so that the art acts to perform the work through its doctrine. Now the question is whether the doctrine of the physicians of the universities is the art of medicine, or whether mine is: this will be proved through works. Now take note what Christ charges in our philosophy— and it is necessary for us to understand it—Christ who not only renewed the eternal light among mortal men, but also natural light, saying: False prophets and false Christs will arise, etc., and will give and perform many signs; thus, although false physicians too perform signs, as they may contend, these are nevertheless no argument against true medicine. For just as Moses and the *Malefici* [14] were ranged against one another in their works, so too the true and the false foundation of medicine. So now I give a true indication how to recognise my *Adversam partem* and myself in the works, and works are found among the false too, as Christ says *De Prodigiis et Signis*. So I will differentiate between them for you in this manner: If there were a man sick of a fever due to end in xii weeks, and then it would be over and done with, and if it happened that the sick man demanded medicine to drive out this fever before its appointed end, he would have two kinds of

[14] I. e. sorcerers.

physician before him, the false and the true. The false proceeds as follows: he begins gradually and slowly to doctor
him, spends much time on Syrups, on Laxatives, on purgatives
and oatmeal mushes, on barley, on pumpkins, on Melons, on
julep and other such rubbish, is slow and frequently administers
enemas, does not know himself what he is doing, and thus
drags along with time and gentle words till he comes to the
term. Then he ascribes the spontaneous end to art. But know
the true physician by this: he divides this *Terminum* [15] into
xii parts, and one and a half he takes for his work, etc.

There is further a great lack of understanding which constrained me mightily to write this work, viz.: they say that the
diseases which I handle in this work, are incurable. Now behold
their great foolishness: how can a physician say that a disease in which death is not, cannot be cured? Thus they say of
Podagra, thus of the falling sickness. Oh! fools that you are,
who bids you speak that can do and know nothing? Why do
you not consider the words of Christ, when he says the sick
have need of the physician? For are not they sick whom you
cast off? I think so. Now if they are sick, as it proves, then
they have need of the physician. Now if they have need of the
physician, why do you say they cannot be helped? It is for
this that they need him, that they may be helped through the
medium of the physician. Why then do they say that they
cannot be helped? They say it because they were born of the
erring of medicine and want of understanding is the mother
who bore them. Each disease has its own physic. For God
desires wondrously to be seen in the sick, for instance in the
diseases of the falling sickness, in stroke, in St. Vitus' dance,
in all others which it is not necessary to mention here. For
God it is, who commanded: Thou shalt love thy neighbor as
thyself and thou shalt love God above all things. If, now, thou
wilt love God, thou must also love His works. If thou wilt
love thy neighbour, thou must not say: For thee there is no
help. But thou must say: I cannot do it and I understand it

[15] I. e. the predestined period.

not. This truth shields thee from the curse that descends on the false. So take heed what is told thee; the rest shall be sought after until the art is found from which good works proceed. For if Christ says: *Perscrutamini Scripturas,*[16] why should I not say of this: *Perscrutamini Naturas Rerum?* [17]

Thus have I wished to defend myself that I justly bring forth and reveal a new medicine according to the present Monarchy. And although it be said: Who teaches thee so to do?, I ask thee: Who teaches the day's foliage and grass to grow? For the Same has said: Come unto Me and learn from Me, for I am mild and humble in heart. From Him flows the foundation of truth and what does not proceed from Him is corruption. The devil is *Mille Artifex* in whom abide many false *Signa* and *Prodigia,* who rests not, who pursues us like a growling lion to make liars of us as well as himself.

You should not be astonished that at the close of this defence I draw your attention and point to Him who said: ' I am meek and humble in heart,' that you should learn medicine from Him Who alone is a teacher of the eternal. But what is there in us mortals that reaches us not and comes not to us from God? He Who teaches the eternal, teaches also mortal things, for they have a common origin. And although it is true that the eternal teaching has spoken in words, and medicine has not, yet when He says: ' The sick have need of a physician,' and ' The physician is of God,' how can the physician not acknowledge Him as his teacher, of Whom he comes? The physician is he who in the bodily diseases takes the place of God and administers for Him, and therefore he must have from God that of which he is capable. For in the same way as physic is not of the physician, but of God, so is the physician's art not of the physician, but of God. However, there are three kinds of physician: one kind is born of nature, delivered by the heavenly physicians, conceived under the influence of the constellation, just as the *Musici* and *Mechanici,* the *Rethorici*

---

[16] I. e. investigate the Scriptures.
[17] I. e. investigate the nature of things.

and the *Artes* are born. Then too another kind: these are the physicians who are taught by men, brought up in medicine and instructed in it, as much as it is possible for men to learn, or according to their abilities. The third is a kind given by God and these have learned of God. As Christ says: Every scribe will be taught by God: that is to say, what we can do we have from God. Now as medicine produces its *Professores* in three ways, should not one see if they do not agree in their Theory and Concepts? In the work they all meet and determine an end and *Terminum*. Nature gives her kind as conception received her influence: man too teaches according to his abilities. God teaches according to His will. But this is the conclusion of all these things: that, the man who desires to teach men must derive his learning from God and from nature, and from the same must men learn. Whatever teaching comes otherwise than from the source, is as is indicated in the next error.

THE OTHER DEFENCE CONCERNING THE NEW DISEASES AND *Nomina* OF THE ABOVE-MENTIONED DOCTOR THEOPHRASTUS

To defend and protect myself, to shelter myself, in that I describe and depict new diseases never before described, and new *Nomina* never before employed, but given by me: Pay heed therefore why this happens—it is because of new diseases, that I may indicate them. I write of the crazy dance which the common man calls St. Vitus' dance, also of those who kill themselves, also of false maladies which befall through sorcery, as well as of people possessed. It seems unjust to me that these diseases should never have been described by medicine, that they should have been forgotten. But what causes me to do this and brings me to it is the fact that astronomy, which heretofore has never been taken up by physicians, teaches me to recognise such diseases. If the other physicians had been as experienced in astronomy, they would have been explained and discovered completely long before me. Since, however, *Astronomia* is rejected by physicians, these diseases and many others, together with their true causes, can be neither recognized nor understood. Since now the medicine of the other authors does not flow from the spring from which medicine takes its origin, the origin and spring of which I may boast, should I not then have authority to write differently from another writer? To everyone it is given to speak, to advise and to teach, but it is not given to everyone to speak and teach things of strength. For you know that the Gospel too testifies that when Christ taught, He spoke as One who had authority and not as the scribes and hypocrites. Such authority one should respect as proves itself with works, if one is incredulous of the word. Therefore am I well aware that little as a man can describe in detail how a thing is formed, if he has never seen it with his eyes, compared to a man who has seen it with his eyes, so the same judgment will be passed here between those who speak without reason and those who speak with reason. It is not less so with a sick man; he belongs under the physician, and justly all diseases should be known to them. However, what the physician may not know in one, that he should know in another. For thus too were the talents of the apostles distributed and what is given to each, in that he takes pride: what is not given to him is no disgrace to him. For as God desires each man to be, so he remains. The other authors cannot boast of such talents: they rejoice in their term, and what they cannot accomplish through the *Terminum* they say is impossible to cure.

Further, that I defend myself because I write new *Nomina* and new *Recepta*: at this you should not marvel. It happens not because of my simplicity or ignorance; on the contrary, everyone can well realise that any simple scholar can read such *Nomina* as were given by the Ancients, also their prescriptions, from the paper and recognise them. But what drives me away from them, is that the *Nomina* are put together and composed of so many different languages, that we can nevermore get things of this kind thoroughly into our heads. And they themselves do not understand and recognise their own *Nomina*; moreover in the German tongue the *Nomina* are changed from one village to another. And although some people have written *Pandectas* [18] and so on, they have hit upon other things which to believe is not in me. And this for many reasons. That I should wish to put myself in such danger and should willingly enter upon such an uncertified apprenticeship, this my conscience will not do. For in these same authors no chapter is free of lies and great errors, but something is found there that spoils it all. How then should these authors please me? I seek not *Rhetoricam* or Latin in them, but I seek medicine of which they can give me no account. Thus too with the prescriptions, they say I write them new receipts and introduce a new procedure. As they have told me to my face: I am to use nothing strange, according to the meaning of God's tenth commandment: Thou shalt not covet strange things. Since now they blame and scold me as a transgressor against the tenth commandment, it is necessary for me to discover what is strange or not strange. Namely, that a man enter not by the right door, is strange; that a man take what belongs not to him. For example, that a man claims to be a physician and is it not, that a man doctor with things in which there is no medicine. Should I be blamed for being able to discover these tricks?

Further, that I should write about people possessed is thoroughly distasteful to them, yet I do it not for this reason. Since fasting and prayer drive out evil spirits, I consider it especially to be recommended to the physician first to seek the kingdom of God and afterwards what he needs will be given unto him. If it is given to him through prayer to make the sick healthy, allow that it is a good purgation. If it is given to him through fasting, allow that it is a good *Confortativum*. [19] Tell me one thing: Is medicine only in herbs, wood and stones, and not in words? Then I will tell you what words are. What is it that words cannot do? As the disease is, so also is the medicine: if the disease is entrusted to the herbs, it will be healed by means of herbs. If it is under the stones, it will also be

[18] In the fourteenth century, for instance, Matthaeus Sylvaticus wrote "Pandectae Medicinae," an alphabetically arranged reference book.

[19] I. e. a strengthening medicine.

nourished under them; if it is subject to fasting, it must be driven away by fasting. Possession is the great disease. Now as Christ demonstrates the remedy, why should not I investigate the same writing, as to what the prescriptions contain and are in this malady? Heaven makes disease: the physician drives it away again. Now just as heaven has to yield to the physician, so too must the devil yield through the right ordering of medicine. The neoteric and modern physicians carry on as they do, because the loquacious Mesue [20] did not think of such things, nor others whose *Aemuli* they are.

I have been charged too with giving diseases new *Nomina* which no one recognises nor understands. Why do I not adhere to the old Names? How can I use the old *Nomina* when they are not derived from the origin from which the disease rises, but are only Sur-*nomina,* of which no one truly knows whether he rightly calls the disease by that name or not. Since then I find and recognise such uncertain ground, why should I give myself so much trouble on account of the *Nomina*? If I understand and recognise the disease, I can well create a name myself for the child. Why should I say *Apoplexis* or *Apoplexia?* or, why should I bother whether *Paralysis* is a product or a corruption,[21] or whether *Caducus fulguris* is called *Epilentia* or *Epilepsia?* or what care I whether it be *Graecum, Arabicum* or *Algoicum?* [22] I am concerned only with finding out the origin of a disease and its treatment and with relating the name to this. Those are things which only waste time with useless talk.

To instruct you further as to the new diseases which I announce in one way or another: there are yet more causes which constrain to seek new diseases. For instance, the sky has a different action every day, changes daily its constitution. The reason is, it too is growing older. For in the same way, when a child is born it changes according to its age, the further the more different from youth, down to the *Terminum* of death.

[20] Paracelsus probably refers to a medieval pharmacological work supposedly written by one Mesue.

[21] "corrumpiert," Byrckmann, p. 16.

[22] I. e. the dialect of the Allgaeu in Bavaria.

Now the sky too was a child, it too had its beginning and is predestined to its end, like man, and death is in it and around it. Now as everything changes with age, the works of the same change also. And if there are changes in the works, of what avail to me then is the rod of young children? For this reason I speak of the present monarchies, because of the age of the firmament and the elements. And further, there is such a throng [23] of people and such an intermingling amongst them with all the conduct of men in fleshly lusts, as have never been before as long as the world has stood. Thence there comes such a *Pressura gentium*,[24] such, too, as has never been before: and from this there results a medicine which has never before existed. Therefore can the physician not make shift as does he who says: I manage with the books which were written two thousand years ago. There are never again the same *Causae*: things are more biting now, as both philosophies of heaven and the elements sufficiently prove. The above-mentioned *Doctores* of medicine should consider better what they plainly see, that for instance an unlettered peasant heals more than all of them with all their books and red gowns. And if those gentlemen in their red caps were to hear what was the cause, they would sit in a sack full of ashes as did they in Nineveh. Thus I know now that in accordance with this Defence and for the reasons given, I can well describe and give new *Nomina* and new diseases.

THE THIRD DEFENCE

CONCERNING THE DESCRIPTION OF THE NEW RECEIPTS

But above and beyond what has been said, the outcry is still greater among the ununderstanding, supposed and fictional physicians who say that the prescriptions which I write are poison, corrosive, and an extraction of all that is evil and poisonous in nature. To such a contention and outcry my first

[23] "menge," Sudhoff (1928), p. 136. Byrckmann and Huser both have "meinung."
[24] I. e. affliction of the people.

question, if they were clever enough to answer it, would be, whether they knew what was poison and what not, or whether in poison there is no *Mysterium* of nature? For in this point they are lacking in understanding and ignorant of natural forces. For what has God created that is not blessed with a great gift for the good of man? Why then should poison be rejected and despised, if not the poison but nature be sought? I will give you an example that you may understand my intention. Behold the toad, how poisonous indeed and detestable a creature it is: behold also the great *Mysterium* which is in it concerning the pestilence. If then the *Mysterium* should be despised because of the poisonous and detestable character of the toad, what a mockery that would be! Who is it who composed the receipt of nature? Was it not God? Why should I despise His *Compositum,* even though what He mixes seems inadequate to me? It is He in Whose hand abideth all wisdom and He knows where to put each *Mysterium.* Why then should I marvel or let myself be frightened because one part is poison, and despise the other part too? Each thing should be used for what it is ordained and we should have no further fear of it. For God Himself is the physician and the medicine. And every physician should acknowledge the strength of God which Christ interprets to us, saying: And though you drink poison, it will not harm you. Now if the poison conquers not but enters without harm when we use it according to nature's ordered way, why then should poison be despised? Who despises poison, knows not what is in the poison. For the *Arcanum* which is in the poison is so blessed, that the poison detracts nothing from it, nor harms it. Not that I would wish to have satisfied you with this verse and Paragraph, or to have defended myself sufficiently; rather is it necessary to give you further account, if I am sufficiently to explain poison.

How is it that you see in me that with which you are all filled and rebuke me for a lentil, when melons lie in you? You rebuke me for my prescriptions: consider yours, how they are. First, for instance, with your purging. Where in all your books is a *Purgatio* that is not poison, or serves not death, or can be used without annoyance, if *Dosis* is not given its proper weight? Now pay heed to what this means: it is not too much, nor too little. He who strikes the middle, receives no poison. And even if I used poison, which you cannot prove, but if I did indeed use it and gave its *Dosin,* am I punishable for this, or not? This I desire to make known to all and sundry. You know that Thyriac is made from the snake *Thyro*: why do you not condemn your thyriac also, since the poison of this snake is in it? But since you see that it is useful, and not harmful, you are silent. If then my medicine is found to be not less than thyriac, why should it pay for being new? Why should it not be just as good as an old one? If you wish justly to explain each poison, what is there that is not poison? All things are poison, and nothing is without poison: the *Dosis* alone makes a thing not poison. For example, every food and every drink, if taken beyond its Dose, is poison: the result proves it. I admit also that poison is poison: that it should, however, therefore be rejected, is impossible. Now since nothing exists which is not poison, why do you correct? Only in order that the poison may do no harm. If I too have corrected in like manner, why then do you punish me? You know that *Argentum vivum* is nothing but poison and daily experience proves it. Now you have this in use, you anoint patients with it, much more thickly than a cobbler anoints leather with grease. You fumigate with its cinnabar, you wash with its sublimate and do not wish people to say it is poison; yet it is poison and you introduce such poison into man. And you say it is healthful and good, it is corrected with white lead, just as if it were not poison. Take them to Nuernberg for examination, the *Recepta* that you and I write, and see there who uses poison or not. For you know not the correction of Mercury, nor its *Dosin;* but you anoint with as much as will go in. One thing I must ask you to think out: whether your *Recepta,* which you say are without poison, can cure the *Caducum,*[25] or not, or *Podagram* or *Apoplexiam;* or whether with your sugar of roses you

---

[25] I. e. epilepsy.

can cure St. Vitus' dance and the *Lunaticos,* or other similar
diseases? Indeed you have not done it with that and will still
not do it with that. So it must be something else: why then
must you be vexed with me when I take what I must and should
take, for what it is ordained? I let Him answer for it, Who
composed it thus in the creation of heaven and earth. Moreover, as the art is given us of separating two antagonistic
things from one another, why then should poison be said from
the beginning to be present? Consider all my *Recepta,* whether
it is not my first principle that the good be separated from the
bad? Is not this separation my correction? Should I not
administer and use such a corrected *Arcanum,* since I can find
nothing evil in it, and you much less? You object to my Vitriolum
in which there is great mystery and more avail in it than in all
the apothecaries' boxes. You cannot say it is poison; if you
say it is corrosive, tell me in what form? You must make it
so, or else it is not corrosive. If it can be made into a corrosive,
you can also prepare it as a *Dulcedinem,*[26] for these are both
together. As the preparation is, so too is the vitriol; and what
every *Simplex* is in itself, is by art made into many beings, into
all shapes and forms like food which stands on the table. If
man eats it, it becomes human flesh, through a dog dog's flesh,
through a cat cat's flesh. Thus is it with medicine: it becomes
what you make of it. If it is possible to make evil out of good, it
is also possible to make good out of evil. No one should condemn a thing who knows not its transmutation and who knows
not what separation does. Though a thing is poison, it may
well be turned into non-poison. Take an example from Arsenic,
which is one of the chief poisons and a *Drachma* kills any steed:
burn it with *Sal nitri* and it is no longer poison. Eating ten
pounds is harmless. Behold then what the difference is and
what preparation does.

But he who would punish, the same should first learn so that,
when he punishes, he does not disgrace himself. I can well
see your foolishness and simplicity in the fact, too, that you

[26] I. e. sweetness.

know not what you say and that one must make considerable allowance for your good-for-nothing tongues. I write new *Recepta,* for the old are useless. And there are new diseases, demanding new *Recepta.* But mark this in all my prescriptions: whatever I may take, I take that in which is the *Arcanum* against the disease which I am fighting. And notice further what I do to it: I separate what is not *Arcanum* from what is *Arcanum* and give to the Arcanum its right *Dosin.* Now I know that I have defended my *Recepta* well, and that you scold me concerning them out of your envious hearts, and offer your useless *Recepta.* If you were of good conscience you would give up: but your hearts are full and your mouths run over. I put five *Defensiones* in this work: read them through and you will find the reasons why I make the *Recepta* from the same Simples as you denounce as being poison. Why should I pay because I lay the foundation which you cannot see? If you were experienced in the things in which a physician should be experienced, you would change your minds. But this you should remember: that what turns out to the good of man is not poison. Poison is alone what turns out to the harm of man, what is not of service to him but injurious, as your *Recepta* sufficiently testify, where no art is considered, except only pounding, mixing and pouring. Herewith then I have desired to defend and protect myself, that my *Recepta* are administered and applied according to the order of nature, and that you yourselves know not what you say, but use your tongues like a madman, uncomprehending and unthinking.

THE FOURTH DEFENCE

CONCERNING MY JOURNEYINGS

It is necessary that I answer for my journeyings and for the fact that I am resident nowhere. Now how can I be against that or force that, which to force is not possible for me? Or, what can I give to or take from predestination? Yet I have to exonerate myself in some measure to you, since I am so much

harangued to, to vex me and ridicule me too, because I am
a wayfarer and as though I were therefore the less worthy.
No one should take it amiss, if I should complain about this.
The journeys which I have thus far made have profited me
much, for the reason that no man's master is in his home and
none has his teacher in the chimney-corner. Thus the arts are
not all confined within one's fatherland, but they are distributed
over the whole world. Not that they are in one man alone,
or in one place: on the contrary, they must be gathered
together, sought out and captured, where they are. The whole
firmament testifies with me that the *Inclinationes* are distributed severally, not only in each person's own village; rather,
according to the contents of the uppermost sphere, the *Radii*
too find their goal.[27] Is it not just and very fitting that I should
investigate and seek out these goals, and see what is wrought
in each one? If I failed in this, I should not rightly be the
Theophrastus I am. Is it not true, art pursues no man, but
must be pursued? Therefore do I have authority and reason
to seek her, and not she me. Take an example: If we would
go to God, we must go to Him, for He says: Come to me.
Now since this is so, we must go after what we want. Thus
it follows: if a man desire to see a person, to see a country,
to see a city, to know these same places and customs, the nature
of heaven and the elements, he must go after them. For, for
them to go after him, is not possible. Thus the way for anyone who would see and experience something is that he go
after the same and competently enquire; and when things go
best, move on to further experiences.

How can a good *Cosmographus* grow in the chimney-corner,
or a *Geographus*? Does not sight give the eyes a true foundation? Then let the foundation be made firm. What then says
the baker of pears in the chimney-corner? What can the car-
penter say without his sight's information? Or what can be
testified without sight? Did not God allow Himself to be seen
with the eyes, and does He not call us to witness that our eyes

---

[27] For this concept cf. Von Waltershausen, l. c., p. 58 ff.

have seen Him? How then should an art or anything else renounce the testimony of the eyes? I have sometime heard of those experienced in laws, how they wrote in their laws that a physician should be a wayfarer: this pleases me greatly. The reason is that diseases wander hither and thither throughout the breadth of the world, and stay not in one place. If a man wish to recognise many diseases, let him travel: if he travel far, his experience will be great and he will learn to recognise many things. And if it were to come to pass that he return to his mother's bosom, if then such a foreign guest come into his native land, he will know him. But if he were not to know him, it would be a mockery and a great shame, for he would be unable to do for his neighbour what he has loudly boasted of knowing. If then what I do for the common good should be taken amiss, it would trouble me. Yet only the cushion-sitters do so, who without a sledge, carriage, and wagon cannot go outside the gates and know not with their art how to get to a shoe-maker's for a pair of shoes except on an ass and for a ducat. If without a ducat thou canst do nothing for a pair of shoes, then art thou thyself an ass and ducat. They are not *Perambulani* [28] either, wherefore they hate that which they are not. They hate the better because they are themselves the worse. Now I know that travel does not spoil but [29] improves. Does not travel make every action better, does not travel confer more understanding than sitting by the stove? A physician should not be a niggard: he should let himself be seen further afield. Nevertheless, fashioned as they are in the world in my times, they like neither to travel nor to learn. To this the people bring them by continually giving them more money, although they know nothing. When they notice that the peasants know not how a physician should be, they stay in the chimney-corner, seat themselves in the midst of books, and ride thus in the Ship of Fools.

A physician should first of all be an *Astronomus*. Now

---

[28] I. e. wanderers. Byrckmann, p. 28, has " parabolani," which makes equally good sense, cf. footnote 30.

[29] " sonder," Byrckmann, p. 28.

necessity demands that his eyes should give him evidence, in order that he may be such: without this evidence he is only an astronomical gossip. It demands too that he should be a *Cosmographus*: not to describe how the countries wear their trousers, but to attack more bravely what diseases they have. Although it be thine intention and desire to be able to make the costumes of this land from what thou hast learned here, and thou excusest thyself thus from gaining knowledge of strange lands, what concern is it of the physician's that thou art a tailor? Wherefore, as the things now mentioned must be experienced, they belong to us *Parabolanis*[30] and are bound up with medicine and not to be separated from it. Thus it is also necessary that the physician should be a *Philosophus* and that his eyes inform him in order that he may be such: if he desires to be one, he must gather together from all quarters what is there. For if a man but desire to eat a roast, the meat comes from one land, the salt from another land, the food[30a] from another land. If these things have to travel till they reach thee, thou too must travel till thou attainest what cannot come to thee. For the arts have no feet for the butchers to drive them after thee; neither can they be put into vats, nor can they be nailed up in a barrel. Now as they have this infirmity, thou must do what they should do. English *Humores* are not Hungarian, nor the Neapolitan Prussian, wherefore must thou go where they are; and the more thou seekest them out and the more thou learnest of them, the greater is thine understanding in thy native land. Thus it is also necessary that the physician be an alchemist: if now he desire to be one, he must see the mother from whom grow the *Mineralia*. Now the mountains go not after him, rather must he go after them. Now where the *Mineralia* lie, there are the artists: if a man wish to seek out artists in the analysis and preparation of nature, he must seek them in the place where the *Mineralia*

---

[30] " Parabolani " originally meant reckless fellows. At the time of Paracelsus, the term was applied to those who, regardless of their own danger, went out to tend the sick, particularly in times of pestilence. Cf. Du Cange, *Glossarium mediae et infimae Latinitatis* . . . VI, Niort, 1886, p. 155.

[30a] " die Speiss," Huser, p. 176.

are. How then can a man find out the preparation of nature, if he does not seek it out where it is? Should I then be blamed because I have wandered among my *Mineralia* and learned the temper and heart, and grasped in my hands the art of those who teach me to separate the pure from the dross, by which I anticipated much evil? Nonetheless I must also repeat the philosophic saying, that wisdom is despised only by the ignorant; thus, too, art by those who do not profess it.

I say nothing of other matters: that he who goes hither and thither makes the acquaintance of many people, experiences all kinds of behaviour and customs that another would eat his hat and shoes to see. I say nothing of greater things than these. Does not a lover go far to see a beautiful woman? How much further for a beautiful art? Now did not the queen come from the end of the sea to Solomon only to hear his wisdom? Now if such a queen pursued the wisdom of Solomon, what was the reason? It is this: that wisdom is a gift of God. Where he gives it, there should one seek it out. Thus too where he has placed art, there should it be sought. It shows great perception in man that man is reasonable enough to seek the gifts of God where they lie, and understands that we are obliged to go after them. If then there is an obligation here, how can one despise or spit upon a man who carries it out? It is true, those who do not thus, have more than those who do: those who sit in the chimney-corner eat partridges and those who pursue the arts eat a milk-soup. The corner-trumpeters wear chains and silk: the wanderers can scarcely pay for ticking. Those within the walls have it cold and warm according to their wishes; those in the arts, were it not for a tree, would have no shade. Now he who would serve the belly, he follows me not; he follows those who go in soft raiment, although these are unfit for wandering. For Juvenal has written of them that only he roams happily, who has nothing. Therefore they consider the same saying and, that they may not be murdered, they just remain in the chimney-corner and turn pears. I think it praiseworthy and no shame to have thus far journeyed cheaply. For this I would prove through nature:

He who would explore her, must tread her books with his feet.
Scripture is explored through its letters; but nature from land to
land. Every land is a leaf. Such is the *Codex Naturae;* thus
must her leaves be turned.

THE FIFTH DEFENCE

CONCERNING THE REJECTION OF FALSE PHYSICIANS AND FALSE COMPANY

As nothing is so pure as not to be defiled with flaws, it is
necessary that one should make known the defiled and the pure,
for it is proved also in medicine that there is more evil than
good. As Christ, however, had twelve disciples and one of
them was a traitor, how much more credible is it among men
that of twelve scarcely one is good? The reason is that while
we should do all things for love, yet nothing is done for love
but only for the sake of squaring and payment, from which
comes selfish gain of which false physicians are born to medicine, so that they seek money and admit not the commandment
of love. Now when a thing has selfish gain for its aim, the
arts are falsified, and the work too; for art and craft must
come from love, else there is no perfection. For in the same
way we have two kinds of apostle; the one loves Christ for
his own gain, wherefore was the purse of selfish gain given
to him. Therefore he had his reason, for his selfish gain to
sell Christ himself, and for his selfish gain to send Him to
His death. If now Christ had to suffer Himself to be sold and
betrayed for the sake of selfish gain, how much more do the
false physicians make man lame and crooked, strangle and kill
him, in order that their profit be increased and not hindered.
For as soon as the love of one's neighbour grows cold, it can
bear him good fruit no more; and such fruit as is borne is for
selfish gain. Thus should we know that there are two kinds of
physicians: those who act for love, and for profit, and by their
works are they both known. Thus are the true ones known
by their love and the just fails not in his love for his neigh-

bour. The unjust, however, the same as act against the commandment, reap where they have not sown and are as rending
wolves; they reap because they wish to reap, in order that their
profit may be increased, ignoring the commandment to love.

Christ gives examples: how the pearl was bought, how too
the field was bought with the treasure, that is to say that love
is not in the many but in the small. As though he were to say:
If thou art a physician, thy pearl is the patient and he is the
field in which the treasure lies. Now it follows from this that
the physician should sell what he has and heal the sick: thus
acts love of one's neighbour. Where, however, it is not so,
but thou keepest what is thine and takest also from the sick
his, now is scripture nowise followed, wherefore can no art
appear to perfection in medicine. For we must keep before
our eyes how the bag of gain was awarded to Judas and the
other apostles were forbidden to have bags, but to eat what was
put before them. Such offerings as these come from love:
demanding, importuning, begging, are not allowed. For one
thing, what we should receive from our neighbour is left to
love, and is not left to our might. Wherefore it follows from
this that to those who walk in the way of God, perfect works
and fruits grow of their talents which God gave them. Those,
however, who do otherwise than Scripture shows, the same
are surrounded with much lamentation and misery, as well as
those from whom they seek profit. Unless it be that God works
in the neighbour against the art and medicine of the false
physician; otherwise no patient is cured at their hands. Let
no one be astonished that I cannot praise gain in medicine.
For since I know how very ruinous profit is, so that the arts
are by profit falsified and directed only to appearances and
purchase, and that this cannot come to pass without falseness,
—which falseness causes corruption in all things—therefore
should the physician not grow from profit but from love. This
last is without care, cares not what it will eat tomorrow but
considers how the lilies of the field are clothed and the birds
fed, much more the man who walks according to the will of
God.

But since in medicine such useless folk are involved who seek and consider only profit, how can it be that I admonish them to love, or how can it have any effect? I for my part am ashamed of medicine, seeing to what utter deceit it has come. Indeed, there is not a desperate henchman, brothel-keeper or dog-killer but wants to sell his human, or dog dripping for gold and heal all diseases with it: though their conscience proves that they are allowed to cure but one disease among them all. But in consideration of their own profit they accept everything that comes to them. Thus too come all the lazy and profligate rascals into medicine and sell their medicine, whether it makes sense or not. Now he who can fill sacks with gold, he is praised, he is a good physician. Thus the apothecaries too and some barbers take medicine upon themselves, behave and carry on as though it were a woodcart, go into medicine against their own conscience, forget their own souls, if only they become rich, prepare house and home and all that belong in it, and dress it up! They heed not that it came undeservedly into their hands, if only it is there! And it has become a doctoral custom—where scripture sanctions it as right, I know not—that a visit should cost a gulden although it be not earned; and that there be fixed fees for the inspection of urine and other things. That one have compassion with one another and fulfil the commandment to love, such things do not become use or custom. Neither is there any more law, but only grab, grab, whether it makes sense or not. Thus they receive golden chains and golden rings, thus they go in silk raiment and thus display their manifest shame before all the world, which they deem an honour and well suited to a physician. To walk around thus decked out like a picture is an abomination before God.

Is it not just that one should shame oneself of a profession which is used so completely in contradiction to its own nature, by such incapable persons? Although the art in itself is a great treasure of nature, it is not given any consideration by such incapable persons. Thus are many who take up medicine, and each one wishes to use it and not to know it. They are thieves and murderers, they enter not by the right door. Their art

is idle talk and barking, the term supports them, and their
knavery and deceit drive them from one land to another, but
not back again. It happens to them as to a messenger who
brings strange news; wherever he goes, he gives the same
sermon; when he returns, he is heeded no more.

It is a grave and pitiful thing, that such an art should be full
of such incapable, careless persons and be thus rendered false,
so that one believes not the truth in it; and that it should have
come to the point when their knavery is so utterly manifest,
that none of us is of good report but all of us are thought
alike, —which in some ways I cannot take amiss. The reason
is: Since the Jews, as a worthless deceitful people,[30b] employ
medicine and are supported by the pharisaic, who then should
think anything of a profession which is controlled by such
knaves? And as long as one wishes to ride all horses with one
saddle and recognises not disease in its essence, rather, what
comes into each man's head, is his art, there is yet no experience nor truth established. But the reason why such things
happen is this: the world desires to be deceived. Therefore
must medicine be manned by such knaves by whom the world
may be deceived, for a pious man does it not. But if the world
did not wish to be deceived, medicine would be manned by
others. But as the world is in some ways of little or no account,
she cannot tolerate the pious in her presence; therefore must
like things be arranged with like. Is it not just that a man
should be ashamed to be numbered and named among such
knaves? Not only that they rummage around in medicine but
also that they are content to display their arrogance; thus they
claim to know and to understand all *Religiones*; they claim
authority to punish or to praise all things; they boast of knowing all languages and when one examines it, it is sealed with
bunkum. One says: the heavens effect such things and the
firmament is the cause. I too know something of the firmament, but I cannot find in it that the falseness in medicine is
born of the firmament. But this I know well, that man's

[30b] Cf. Introduction, p. 7.

carelessness is a cause of deceit and one needs accuse no one
of it but oneself. No longer is anyone willing to journey to
become a master. Each wants to fly before his wings have
grown. This is the deceit: everyone acts, but knows not what.
This is the carelessness that is in man, that he undertakes a
work knowing that he cannot do it. So, however, the false
physician thinks: If it succeed not, (as will indeed happen), thou
canst well vindicate thyself and defend thy knavery with God,
or lay the blame on the patient: thus they must give thee money,
whatever happens. Medicine is an art which should be employed
with much conscientiousness and much experience, also with
much fear of God; for who fears not God, murders and
steals for ever and ever; who has no conscience, has no shame
in him either. It is a wicked shame, or perhaps a calamity,
not to recognise such godless people and to cut down a tree
which is of no use, and throw it into the fire. For this is how
they are: since they see the leniency of the authorities and see
too that they sometimes love profit, they are after it like whores
by the moat. Therefore is it necessary that a distinction be
made between the physicians who walk according to the law
of God and those who walk according to the law of man. The
one serves in love, the other for profit. Thus I desire to have
defended myself in this place, that I have nothing in common
with the Pseudo-physicians, nor do I favour them, but would
further the cause of felling the tree. It need not for my part
be long delayed.

The Sixth Defence

To Excuse His Strange Manner and Wrathful Ways

Not enough to attack me in various articles, but I am said to
be a strange fellow with an uncivil answer, I do not wash up
to the satisfaction of everyone, I do not answer everyone's
contention in humility. This they consider and deem a great
vice in me. I myself, however, deem it a great virtue and would
not that it were otherwise than it is. I like my ways well
enough. In order, however, that I may justify myself as to

how my strange manner is to be understood, pay heed: I am
by nature not subtly spun, neither is it usual in my country to
attain anything by spinning silk. Neither are we raised on
figs, nor on mead, nor on wheaten bread, but on cheese, milk
and oatcakes. This cannot make subtle fellows; besides what
one received in youth sticks to one all one's days. The same
is almost coarse to the subtle, the cat-clean, the superfine. For
those who are brought up in soft raiment and in the women's
apartments, and we who grow up among fir-cones do not
understand one another well. Therefore must the coarse be
judged coarse, though the same think himself utterly subtle
and charming. Thus it is with me too: what I think is silk,
the others call ticking and coarse cloth.

But pay heed further how I justify myself in this accusation that I give a rough answer. The other physicians know
little of the arts; they resort to friendly, pleasing, charming
words; they advise people with breeding and fine words; they
set forth all things at length, delightfully, with distinct differ-
entiations, and say: Come again soon, my dear sir; my dear
wife, go and accompany the gentleman, etc. I say thus: What
wilt thou? I have no time now; it is not so urgent. Now I
have upset the applecart![31] They have made such fools of the
patients that they are completely of the belief that a friendly,
affectionate manner, ceremony, ingratiating ways,[32] much ado,
constitute art and medicine. They call him 'young sir' who
only comes from the shopkeeper's; they call another 'Sir,
wise Sir' who is a cobbler and a dullard, where I say 'Thou';
but with this I throw away my resources. My intention is to
gain nothing with my tongue, but only with works. As they,
however, are not of this opinion, they can well say in their
way that I am a strange, queer-headed fellow, that I give little
good advice. I do not believe in feeding myself on friendly
caresses, wherefore I cannot use what befits me not, nor what
I have not learned. For it is not necessary to use such flattery

---

[31] Huser, p. 183: "Jetzt hab ich in den Pfeffer gehofiert," which defies
literal translation.
[32] Cf. Sudhoff (1915), p. 37, footnote.

and to deal tenderly with every boor [33] who is not fit to be carried in a dung-barrow. Medicine should be such, that the physician may answer according to his flesh and blood, his country's customs and his own nature: rough, rude, stern, gentle, mild, virtuous, friendly, delightful—according to how he is by nature and by acquired habit. But let this not be his art, but only the briefest answer. And on with the works! That's the way to oil the wheels! [34]

Thus I consider in this respect I am sufficiently defended. Still, it happens that I have other strange ways, for instance towards the sick, if they do not follow my prearranged injunctions. Anyone can judge that such strange ways are not unjustified, in order that medicine may be found true, the patient become well, and I may still remain without blame. A turtle-dove would grow angry with such lousy muddle-heads.

There is a further complaint against me, with regard in part to the servants who have left me, in part also to my Pupils, that none of them could stay with me because of my strange ways. Now note my answer: the hangman has taken from me into his favour one and twenty servants, and removed them from this world, God help them all. How can a man stay with me, when the hangman will not leave him with me? And what has my strange way done to them? If they had fled the hangman's way, that would have been true art. And there are yet some who have thus kept with me and have also fled the hangman; and in rejecting me, they have excused themselves, for I was strange, no one could cope with me. But how should I not be strange, when a servant is no servant, but a master? He looks for his advantage, ruins me with it, brings shame upon me, and rejoices in it. Thus they lie about me to the patient, they receive them behind my back, without my knowledge and consent, make a contract with them for half the money, say they know my art, have copied it from me.

[33] Cf. ibid., p. 38, footnote.
[34] Huser, p. 184: "das heist dem Rappen Muss in das maul gestrichen." The phrase cannot be rendered literally. For meaning, cf. Schweizerisches Idiotikon, VI, 1909, col. 1169.

After such disloyalty they cannot and would not be with me, nor the patients either. Afterwards, when I hear of it, the knavery is a bargain. Let anyone judge how honest the bargain is! *Doctores,* barbers, bath-keepers, pupils, servants and lads too, have done it to me. Should this make a lamb of me? A wolf should result in the end! Moreover, I must go afoot while they ride. For this comforts me always, that I linger and remain when they run away and their falsity is understood. Not less do the *Doctores* complain about me and not without justification. For telling the truth hurts the one whose cunning is revealed. How many, however, are there who for this speak good of me? —they too are *Doctores.* Thus are the apothecaries, too, inimical to me; they say I am strange, eccentric, etc., no one can do right in my opinion. Yet to me, everyone can do right who acts righteously. But as for giving *Quid pro Quo,*[35] *Merdum pro Musco*[36]—it suits me not to acknowledge the *Quid pro Quo* book of the travelling scholars, to accept it nor yet to allow its use. Besides, of what they give me themselves, not a third is good, sometimes indeed none is good. And the same is sometimes not what they say it is. Should I subject my patient to the *Quid pro Quo,* even if it is no good? Thus should I come to shame, my patients to disaster, perhaps even to their death. When I proclaim this in my native way, which I consider and deem very friendly, the idlers [37] call it an angry, strange manner. Other *Doctores* do not thus, I alone do so. And furthermore, I write short prescriptions, not forty to sixty ingredients, I prescribe little and seldom, I do not empty their boxes for them, I do not bring much money into their kitchens. Yet this is the business for which they calumniate me. Now judge for yourselves, to whom do I owe most? Or to whom have I pledged myself as *Doctor*? To help the apothecary empty the bags in his kitchen, or the patient from the kitchen to his benefit? Now behold, dear sirs, how strange I am and in what a plight is my head! If I should defend my

[35] Cf. Introduction, p. 7.
[36] I. e. dung for musk.
[37] Cf. Sudhoff (1915), p. 39, footnote.

angry manner to the end, they would blush with shame and be frowned upon. For to tell of those who thus accuse me and criticise me because they think thus to belittle me, will bring too much of their knavery to light for them, and will greatly damage them with all pious judges and interrogators. If I were now to attack some barbers and bathkeepers a little and make public what they have against me, why they call me eccentric and a strange man, I rather think there would be few of them and they would fare very much as some others have fared whom I have mentioned. Wherefore, understand me in this sixth defence, that you who hear such things may please to measure them with true judgment and true balance; and remember that not everything comes of a pure heart, but from filth with which their mouths run over, to glorify themselves and to belittle me.

The Seventh Defence

How I Too Know Not All, Cannot and Am Not Able to Do What Each One Needs, or Might Need

This I must confess, that I am not able to grant and fulfil every one's desire, as he certainly and undoubtedly would have it from me: this I cannot do, nor is it in my power. For God did not create medicine thus according to their will, that it might act immediately according to the will of all comers. If then God will not grant nor give such people anything, how can I help it? For I cannot master nor overpower God, but He me and all others. Thus are all answered alike: if they were agreeable to God or pleasing to him for healing, He would not have withdrawn nature from them. It is the same thing as one who thinks himself a fine, handsome fellow and wants to stand out before all others and wants all women and maidens to favour him. But he was born crooked, has a hump on his back like a lute and in other respects too he has no comeliness of body. How can women favour one whose own nature does not favour him, but spoiled him in his mother's womb and made nothing good of him? In order, however, that I may instruct you the better, know ye that if God bestow no good on a man, how then should nature bestow good on him? Where both favours are absent, of what good is the physician? And who can blame him? Now they say when I come to a patient, I know not immediately what ails him, but I need time to find it out. It is true. That they judge immediately is the fault of their foolishness; for in the end the first judgment is false and from day to day they know the longer, the less, what it is, and make liars of themselves. Whereas I desire to approach from day to day, the longer, the closer to the truth. For with hidden diseases it is not as with the recognising of colours: in colours one sees well what is black, green, blue, etc. But if there were a curtain before it, thou also wouldst not know. To see through a curtain requires effort where there has been none before. What the eyes see can well be judged hurriedly, but what is hidden from the eyes it is in vain to conceive as though it were visible. Take an example from the miner: however good, true, ingenious, clever, he may be, when he sees an ore for the first time, he knows not what it contains, of what it is capable, how it should be treated, whether to roast it, to smelt it, to refine it, or to burn it. First he has to filter it, try a few tests and experiments, and see what to do. Then when he has sieved [38] it well, he can start on a definite line: that is the way, thus must it be. So is it too in the hidden, tedious diseases: an opinion cannot be formed so quickly— (unless it be by the humorists). For it is not possible to find a dog so quickly, nor a cat in the kitchen; how much less so in a dangerous, secret business? Wherefore, considering, judging, testing things as much as is necessary to the experiment, is not to be taken amiss. Then to it with the true art: that's the fine thing, that's the valuable thing, that's how to treat such diseases. But the humorists do not test by experiment, but by the Reader's enquiries and Proofs. Wherefore many escape to the churchyard before they find out, and even then they do

---

[38] Cf. *ibid.*, p. 41, footnote.

not find out. Thus is their art; and should such an art judge me? I cannot do everything. What can they do, who think nothing is of value but what is cured by the Summa,[39] i. e. their Avicenna, their Rabbi Moises.[40] Short work, whatever the result: these are their *Aphorismi,* which end very abruptly in the churchyard.

Why do you throw it in my face if I cannot cure impossible things, when you cannot cure the possible? But rather you ruin it, so that I must build it up again. How can I cure a cut-off heart, put a cut-off hand back on? Who has ever been able by the light of nature to join death and life and unite them, so that death should receive life? Indeed it is not natural, but divine. How should I do such things when you cannot heal wounds in which there is no death, except what you invite? You are long-sighted, you see into the distance, but not yourselves near by. I will prove it with your conscience, that it teaches and instructs you, that you do and act against it and would glorify yourselves with that which brings you to shame. For you have medicine from God, with which to drive out all things possible; you could do it, and are incapable. Why then do you accuse me of doing nothing for impossible diseases and that for them no medicine is given me, nor created? Know then further the conclusion of this defence. How can I cure possible things when it is hailing in the apothecary's shop? When the storm strikes the kitchen, who can eat of it? How can a fur coat be for shots and armour for the cold? How can I cure with *Quid pro Quo* with which you ruin all your patients? You need luck to be successful with *Quid pro Quo* and to complete the cure. Who can accomplish with false drugs that for which the true only are fitting? Who can complete what he intends shall be done with green herbs, when he is given mouldy ones? Who can bear to be given *Succum Tithymalli* for *Diagridium?* [41] Who can bear or suffer

---

[39] Byrckmann, p. 48. Cf. Sudhoff (1928), p. 412.

[40] I. e. Maimonides.

[41] I. e. the juice of spurge for scammony.

to be given *Picem Calceatorinam distillatam pro oleo bene-dicto*,[42] and cherry mush mixed with Thyriac for a *Mithridatum*? [48] And if I should tell the truth about your *Simplicia* and *Composita,* as necessity requires, where would it lead me?

With this I desire to have defended and protected myself for the last time until further provocation, which, if God will, will also be returned blow for blow.

And herewith too I desire only to beg that the pious and just of true conscience should not bother themselves with my writings. For necessity required that I answer. For Christ too replied and was not silent. For everyone should know that answering is just and fitting, in order that those do not stagnate completely nor blind themselves with idle talk, who live on and enjoy idle talk. If they were not answered, they would win the argument and would consider themselves right, and there would afterwards be even more error, rubbish, disaster and corruption. Wherefore is answering equivalent to anticipation of present and future corruption and a revelation of what the ranters are. For such reasons as these then I have been pleased to answer and to protect myself from all those whose hearts are full of ill-will, that both sides may stand revealed. For it is necessary that vices come; but woe unto him through whom they come! That is to say, it is necessary that liars speak contrary to the truth; but woe unto them, for truth reveals the lies. If they were silent concerning their vices, truth too would be silent. But since it is necessary, lies and vice cannot, and may not, be silent; it must be uttered, but woe unto them. Thou, reader, however, shouldst consider and measure all things most justly, that thy reading may bear thee fruit, profit and good.

---

[42] I. e. distilled cobbler's pitch for *oleum benedictum,* which latter was a complex remedy.

[48] I. e. a complex antidote.

CONCLUSION

And so, Reader, hast thou to some extent understood me in this reply and seen well that I have attacked with all mildness. And by thyself mayest thou well be able to judge what frivolous and useless folk say and do. Thou mayest also remember that all this started only with physicians and then think with what manner of people medicine is supplied, what an unequal pair are Podalirius [44] and Apollo, and now those of today. Would not nature itself perhaps be alarmed at such conditions? For nature knows her enemy very well, as a dog knows a beater of dogs. Holy Scripture sufficiently proves with what praise medicine should be lauded and with what honours the physician. But it stands to reason that there the reference concerned Hippocrates, Apollo and Machaon, who healed with the true spirit of medicine, performed *Prodigia, Signa* and *Opera* and appeared as lights in nature. My simple head can well understand that Holy Writ did not mean those who are without works, the ranters, nor the *Mercenarios,* but those who have trodden in the footsteps of Machaon. It is good to note in the above writings that there is pain and toil on earth. But I consider that if the authorities had thus learned well to recognise these things and had lain sick in the same hospital, there would be more friendly attention to neighbourly love. The laudable province of Carinthia has kept it in mind and represents Maecenas, and gives *Asylum Hippocraticorum* in our times, protecting and sheltering. For this may God give recompense, peace and unity. Amen.

[44] Byrckmann, p. 51.

II

ON THE MINERS' SICKNESS AND OTHER MINERS' DISEASES

BY

THEOPHRASTUS VON HOHENHEIM
CALLED PARACELSUS

TRANSLATED FROM THE GERMAN, WITH AN INTRODUCTION

BY

GEORGE ROSEN

INTRODUCTION

The labouring classes have always been a fundamental factor in the social and economic organization of society. Their work at any particular time has been performed within the framework of an existing productive system. Certain problems, flowing from their functional relationship to their jobs, their communities, and in a larger sense their entire environment, are especially characteristic of the working classes. In this category the problem of occupational disease occupies an important and prominent position. Such diseases are the result of the noxious aspects of industry, which impinge upon the worker in the course of his activities.

The diseases incident to certain occupations attracted attention very early. Mining, as one of the most dangerous of all industries, was one of the first to be investigated with regard to occupational disease. Mining is one of the most fundamental technological disciplines of civilization. Without it our civilization could not exist. The raw materials, which are necessary for even the simplest industrial processes, can be secured in sufficient quantity, only by digging them out of the bowels of the earth. This is not a modern development, however, for evidences of mining operations are to be found among the earliest human cultures. The history of mining extends in unbroken continuity from neolithic times to the present. Just as old, however, is the history of occupational disease among miners. A miner's life has always been a dangerous one. Besides being exposed to various traumatic accidents, his health and life are endangered by the deleterious environment incident to his occupation.

Nevertheless, despite scattered references in the literature of antiquity to this subject, it was not until 1567 that the first monograph, devoted to the occupational diseases of miners, appeared at Dillingen in Germany. The author of this work was Theophrastus Bombast von Hohenheim, usually known as Paracelsus; the book was entitled *Von der Bergsucht und*

45

anderen Bergkrankheiten (On the Miners' Sickness and other Miners' Diseases).

During his life of trials, tribulations and struggles, Paracelsus had ample opportunity to become acquainted with the mining industry of his day. He was instructed in chemical and metallurgical principles by his father. When his father moved from Einsiedeln to Villach in 1502 his opportunities increased. Villach was surrounded by numerous mines and smelters, a fact, which Paracelsus corroborates in his *Chronik des Landes Kaernten.* " Die im Lande Kaernten sind die ersten die in diesen teutschen Landen gewesen, was da angetroffen hat die Metalle, die Vitriole, die Erze u. dgl. Sie sind erstlich in diesem Lande gelernt worden und dann in andere Länder getragen und sind dort Bergwerke nach dem kaerntischem Brauch in das Werk gebracht worden." [1] Paracelsus increased his knowledge of metallurgy in a practical manner. In his work *Von der grossen Wundarzney* he relates that from his 20th to his 25th year he was employed in a smelting plant at Schwaz, in Tyrol. During his journeys through Denmark and Sweden, and later in Meissen and Hungary, he learned about the mines in these countries. Around 1533 he passed through the industrialized Inn valley, which contained many mines, and it is likely, as Sudhoff suggests, that the hygienic conditions in the mines and the occupational diseases of the miners awakened his interest then. Again in 1537 he came into contact with the mining industry, when the management of the Fugger mines called him to Villach to take charge of the metallurgical work there. This evidence is sufficient to show that Paracelsus had enough opportunity to study the mining industry and its workers thoroughly, and to observe the diseases of the miners, particularly the effects of various minerals and metals on the human organism.

The monograph of Paracelsus was probably written around

[1] " Those in the land of Carinthia were the first in these German countries to discover metals, vitriols, ores and the like. The knowledge of these matters first developed here and was then carried into other countries and mines were developed there according to the Carinthian usage."

1533-34, but this is by no means certain. Sudhoff has expressed the opinion that a visit to the mines and smelterworks of the lower Inn valley, in the Tyrol, during 1533, stimulated Paracelsus to write this monograph and that he completed it in 1534. It is possible, however, according to Koelsch, that Paracelsus did not write the book until 1537-38, during which time he was employed in the Fugger mines at Villach. At any rate it did not become known until after its publication in 1567 and its influence did not become apparent until the next century. It was published posthumously—Paracelsus died in 1541—by the editor Samuel Architectus (Zimmermann). Nothing is known concerning the manuscript used by Architectus, but there is no doubt concerning the authenticity of this work.

The monograph, *Von der Bergsucht, etc.* consists of three books, each containing four tractates. The first book deals with the diseases, mainly pulmonary affections, of miners, whereby etiology, pathogenesis, symptomatology and therapy are successively discussed. The second book treats of the diseases of smelter workers and metallurgists, and the third concerns itself with diseases caused by mercury. As this brief outline indicates Paracelsus attempts a monographic description of the peculiar morbid conditions, which afflict mine and smelter workers; this exposition is based upon a consideration of the etiological factors involved. The theoretical conceptions employed by Paracelsus in this work can only be understood within the bounds of his general doctrines. It is very likely, however, that his theories were profoundly influenced by his experiences in, and his knowledge of, the mining industry. Nevertheless, just as Paracelsus and his ideas cannot be understood without an analysis of his background, of Galenic, i. e. Greek and Arabic, medicine on the one hand, and of medieval German mysticism on the other, his account of miners' diseases cannot be fully comprehended without a knowledge of his teaching. His monograph on miners' diseases is a direct application of certain general ideas on etiology and pathogenesis as expressed in his Opus Paramirum. We have here a very interesting example of the interaction of several factors to produce a specific result. Influenced by his mining experience, Paracelsus was led to certain generalized ideas on pathology and etiology. To demonstrate the validity of his teaching, he turned to a definite disease group, and employed his theories to explain the facts that he observed and the therapeutic measures to be employed in the treatment of these maladies. The work that resulted was *Von der Bergsucht.*

Paracelsus considers the miners' sickness (die Bergsucht) to be a disease of the lungs, like "the old and long known lung sickness." "The miners' sickness is . . . the disease of the miners, the smelters, the pitmen and others in the mines. Those who work at washing, in silver or gold ore, in salt ore, in alum and sulphur ore, or in the vitriol boiling, in lead, copper, mixed ores, iron or mercury ores, those who dig such ores succumb to the lung sickness, to consumption of the body, and to stomach ulcers; these are known to be affected by the miners' sickness (bergsuechtig)." Apparently Paracelsus knew nothing of any previous references to the subject of miners' diseases for he states, "that nothing is found about these diseases in the old writers. For this reason they have remained undescribed until now, and the cure too was omitted."

The origin of the disease of the lungs is to be sought in the air, for "the air is the body (corpus) from which the lung receives its sickness." The chief element of the air is the chaos (gas), but this chaos is in turn ruled by the stars. Therefore the physician must concern himself with these sidereal influences, with the constellation of the stars. At the first glance anyone who reads this work, may have the impression that these "constellations" are conceived by the author in an astrologic sense. This view, however, is false, and is soon corrected if the reader is willing to take the trouble of delving somewhat deeper into the ideas of Paracelsus.

Paracelsus regards man as a product of nature. Nature is the macrocosm, man the microcosm, and the former influences the latter. This influence is exerted indirectly by means of the astral constellations since they determine the state of the

" upper firmament," which for Paracelsus includes all kinds of atmospheric conditions. Heat and cold, dryness and moisture are the results of the astral influences and when they act on man they can produce disease. The knowledge of these meteorologic and climatic conditions and of their influence on man is therefore essential for the physician. Paracelsus calls this knowledge astronomy.

While considering these physical factors, Paracelsus also mentions another factor, a chemical one, which plays a rôle in the causation of the miners' sickness, namely the poisonous effects of metals, when they are extracted. He compares the poisonous action of minerals and metals, such as arsenic to the action of the stars in the chaos.

Paracelsus then goes on to discuss the pathological changes in the respiratory organs in cases of miners' sickness. He conceives these changes as being " exudations," of " tartarus." This theory of the " tartarus " diseases was erected by Paracelsus and plays an important part in his medical thought. The " tartarus " concept, however, was not invented by Paracelsus, but was current among those acquainted with alchemistic principles. It had been introduced into Europe from the Orient in the 13th century and is mentioned by Albertus Magnus. Tartarus or tartarum was a general term comprising all forms of precipitation or sedimentation. Paracelsus employed this concept to explain a large group of pathological processes by means of simple physico-chemical principles, such as coagulation, precipitation, sedimentation and deposition. However, the " tartarus " itself was not a simple substance, but a mixture of mercury, sulphur and salt. These three substances are not to be considered as identical with the modern elements and compounds of the same name but rather as the three basic categories of matter, of the macrocosm and microcosm. These categories describe the reactions of substances when exposed to heat; sulphur is that which burns, quicksilver that which evaporates when heated, and salt that which resists heat. There are as many kinds of mercury, sulphur and salt as there are substances.

The second tractate of the first book is concerned with "the origin and birth of the miners' sickness." Paracelsus enumerates the various climatic influences which are harmful to the miners' health. The miners become overheated while working and catch cold easily. The relation between sweet and sour beverages and pulmonary disease as presented by Paracelsus is unclear. He reasons by analogy very often, and Koelsch suggests that he may be thinking of odors in the mine air, due to the mine gases of putrefaction, or more probably of the sweet or sour tastes of various minerals and salts. This view is in all probability correct.

The presence of fat in the chaos also causes the miners' sickness. Fatty vapor is identical with sulphur for Paracelsus, and if this sulphur enters the lungs, it remains there as a resin. The presence of sulphur in the chaos is not the only cause of pulmonary lesions, for there are also mercurial, i. e. vaporous, substances in the air. These arise from the minerals; therefore it is important for the physician to know and to be able to recognize what minerals a particular region possesses. The minerals act either as mercury, sulphur or salt and enter the lung in these forms. The presence of these components varies, however, according to the nature of the minerals present in the different regions. These differences are reflected in the three types of the miners' sickness. Paracelsus again emphasizes the importance for the physician of a knowledge of meteorology, geology and mineralogy. This knowledge of the earth and its constitution he calls philosophy. In his diagnosis, examination and treatment the physician must therefore employ his knowledge of astronomy and philosophy.

In the third tractate Paracelsus discusses the signs and the nature of the miners' sickness, and hints vaguely here at the therapeutic usefulness of chemistry. At this point Paracelsus makes a remark which is very interesting. He reminds the reader that "we must also have gold and silver, also other metals, iron, tin, copper, lead and mercury. If we wish to have these, we must risk both life and body in a struggle with many enemies that oppose us." It would appear from this remark that Paracelsus realized fully that the rise of occupational diseases among the mine workers was a necessary and concomitant result of industrial development.

In his discussion of the miners' sickness Paracelsus now turns to the action of the poisonous vapors, which arise when the ores are roasted, and their effect on the miners. He recognized that the poison can enter the body either as a solid being taken by mouth, or by being inhaled as a vapor. His observations taught him, however, that the action of the poison is much weaker, when inhaled, and consequently the duration and the course of the disease caused by the vapor is different. In other words, he realized clearly the distinction between acute and chronic poisoning. As an illustration of these statements he describes the picture of chronic arsenic poisoning with the characteristic symptoms, pallor, thirst, gastro-intestinal disturbances, and skin eruptions. Finally Paracelsus observes that such miners tend more easily to become victims of other diseases.

In order to facilitate and to simplify the study of these poisons and their effects Paracelsus arranges them in groups. One group comprises the arsenicalia, a second one antimony and similarly acting metals and minerals, while the alkali group is the third one. It is difficult to identify the minerals and ores mentioned under each of these headings and the same is true of the symptoms produced by them. When one considers that these minerals were not pure metals and that the miners were also exposed to other deleterious influences, it is not difficult to realize that neither the classifications nor the symptoms described by Paracelsus can be defined in terms of our present knowledge. Every physician is of course interested in nosographic descriptions and it is difficult to resist the temptation of attempting to identify them in terms of modern nosologic categories. The validity of such identifications is, however, highly hypothetical and always open to question. Some of these symptoms are probably due to diseases of the respiratory organs, either acute lesions, such as pneumonia or pleurisy, or more chronic ones, such as pulmonary fibrosis, tuberculosis, bronchitis with emphysema, or pulmonary neoplasms. The other symptoms may be interpreted as the signs of acute or chronic poisoning, perhaps the cutaneous signs were the result of an occupational dermatitis.

With the mention of the " arcana " Paracelsus indicates and develops his therapeutic theory. The physician who would cure a certain disease should not attempt to find a separate remedy for each new patient because such a method leads only to errors. Instead he must attempt to discover the arcanum which will cure all those kinds of disease exhibiting similar symptoms. This arcanum or remedy can only be found by discovering the cause of the disease, for the cause is also the cure. This theory can also be applied to the miners' sickness for the minerals also have their arcana. Having indicated the theoretical basis of his therapy, Paracelsus devotes the fourth tractate to practical therapeutics.

The tractate opens with an enumeration of the subjects to be discussed—preventive measures, the diet of the miners, the treatment of the miners' sickness, and finally the special treatment by means of the arcana. Having indicated the general measures for the prevention and treatment of miners' sickness, Paracelsus discusses the special therapy of the pulmonary condition. The therapeutic measures which he recommends are based on his physico-chemical " tartarus " theory and on his doctrine of the arcana. The lungs are covered and blocked with " exudates " of tartarus, which can be removed and washed off by profuse sweating.

The second book deals with the diseases of smelter workers, and in the third book Paracelsus concerns himself with the diseases produced by mercury. It is interesting to note that he devotes one third of his monograph to the morbid conditions resulting from mercury poisoning. Mercury was mined and refined in Tyrol, Carinthia and Carniola, regions with which Paracelsus was well acquainted and where he had enough opportunity to study these conditions.

The first tractate of this book is concerned with a description of the characteristics and properties of mercury. The following two tractates deal with the pharmacologic action of the metal on the human organism, which Paracelsus attempts to explain by means of the " astronomical physica." Mercury is a " cold " metal and therefore the mercurius vivus produces shivering, chattering of the teeth and the like. In the fourth tractate Paracelsus attempts to explain the effect of the mercury in the body and how the various symptoms arise as result of this action. The cold or as he also calls it " the winter " that resides in mercury causes " a shivering " without any frost being felt.

It is very difficult to elucidate the nosographic pictures that Paracelsus describes. Clearly not all the things he mentions were due to mercury poisoning. Certain signs and symptoms, however, are definitely due to mercury; such are tremor, gastrointestinal disturbances, oral putrefaction, cachexia and blackening of the teeth.

The monograph closes with a section on the therapy of the mercurial diseases. Paracelsus bases his therapy on the premise that the mercury is deposited in the body, sometimes in such a stable form that it must first be mobilized before it can be excreted. He imagines the deposition of mercury as occurring in a mechanical manner, i. e. the mercury settles into certain spots, preferably dependent parts, by the force of gravity. As soon as any of these spots has been determined, steps must be taken to create an opening through which the mercury can flow out. This is done by applying a corrosive plaster for 2 to 3 weeks and producing an ulcer.

The dead mercury must be revived so that it can pass out of the body. For this purpose Paracelsus recommends baths, either with herbs or with sulphur. It is of interest to note that this form of therapy is still used at the present time. A separate chapter is devoted to the therapy of the oral and dental complaints, such as putrefaction in the mouth (stomatitis), blackening, shakiness, and loss of the teeth and strong, stabbing pains.

Finally mention should be made of a fragment containing two chapters, which was first published in 1589 by Huser in his edition of the collected works of Paracelsus. The fragment is concerned with spirits in mines and their possible harmful effects on the miners. In view of the general acceptance of the existence of such spirits in the mines it occasions no surprise to find that Paracelsus also includes this subject in his monograph. Another reference in this fragment to lightning and thunder in the mines may be interpreted as the expression of the author's experience with mine explosions, but the entire passage is vague so that a definite opinion is not possible.

Paracelsus' monograph was only a beginning but there can be little doubt that it was an important one. Sudhoff has remarked that this work is unique in the literature of the 16th century, and this very fact is one of the major reasons for its significance. It is the first monographic presentation of the diseases of a definite occupational group, the mine and smelter workers. Paracelsus was not content with authority and aprioristic speculation but demanded that nature be investigated. The numerous correct observations that he made are evidence of his own experience in the mines. Paracelsus recognized two large groups of diseases which affected the miners of his time, the diseases of the respiratory organs and the pathologic conditions resulting from the ingestion or inhalation of poisonous metals. He knew that the respiratory diseases were caused by climatic conditions and that they appeared frequently in miners with such symptoms as dyspnea, cough and cachexia. He recognized the poisonous effects of various metals and differentiated acute and chronic poisoning. In the case of mercurialism his description is detailed, most of the important symptoms being mentioned.

Paracelsus realized the importance of prophylaxis and insisted that " the mine spirits must be forestalled so that they will leave the miner uninfected; this is begun, before he becomes subject to the mine and the ore spirits." It is striking, however, that Paracelsus does not refer to any protective apparatus, such as respirators. In this connection it is of interest to note that Paracelsus does not pay any special attention to dust as a causative factor in miners' diseases. It is remotely possible, however, that some of his references to vapors in his treatise may be allusions to dust.

This monograph on the diseases of miners exerted a definite influence on the development of this branch of occupational medicine. For the next 150 years after the appearance of this work every writer on the subject referred to Paracelsus. Some treated his book favorably, others criticized it mercilessly but no writer failed to touch upon it.

The great significance of this monograph lies not only in the circumstance that it is the first comprehensive account of the diseases of any occupational group, but also in the attempt of Paracelsus to relate these diseases to the general body of medical knowledge and theory. In judging his work due regard must be paid to the disadvantages under which he laboured. It should be remembered that Paracelsus lacked any clear cut theoretical concepts and an adequate terminology with which to express his ideas. By writing in a vernacular tongue this difficulty was only increased. Consequently in order to present an adequate theoretical explanation of his observations, he very often reasons by analogy. Parts of his book are therefore rather obscure for the modern reader.

The present translation attempts to represent both the thought and the style of Paracelsus as far as it is possible in English. It is based upon the text in Sudhoff's edition of the collected works of Paracelsus (Abt. I, Band IX). Several other works were consulted to elucidate obscurities in the text. The commentary of Koelsch and Sudhoff's recent biography of Paracelsus were very helpful in this respect. B. Aschner's translation into modern German was also consulted, although it deviates considerably from the original text at times.

ON THE MINERS' SICKNESS AND OTHER MINERS' DISEASES

THREE BOOKS

by the very experienced philosopher and physician

PHILIP THEOPHRASTUS VON HOHENHEIM

THE FIRST BOOK, CONTAINING FOUR TRACTATES

Of the first tractate, the first chapter.

In order to describe the disease of miners' sickness, it is first necessary to uncover the old and long known lung sickness, and how this occurs. The birth and origin of the miners' sickness are to be understood in a similar manner. These two diseases differ only in the element and in the location, but their course is the same. Now the lung sickness is a disease of the lungs and just as it is possible for it to spread further in the body and to poison it, thus it is also possible for the miners' sickness to do so. In the first book, namely in the first tractate, follows the description of the birth of the lung disease, and how it arises. It is mentioned briefly, as much as is necessary for an understanding. After this has been understood, the second tractate follows in which is described the birth of the miners' sickness. After this knowledge follows the third tractate concerning the appearance and nature of the external and internal body together, in this disease. And in the last tractate the cure of the same miners' sickness, and everything that is necessary for it are described.

In order, however, that you may know what the miners' sickness is, it is the disease of the miners, the smelters, the pitmen and others in the mines. Those who work at washing, in silver or gold ore, in salt ore, in alum and sulphur ore or in the vitriol boiling, in lead, copper, mixed ores, iron or mercury ores, those who dig such ores, succumb to the lung disease, to consumption of the body and to stomach ulcers; these are known to be affected by the miners' sickness (bergsüchtig). Know further that nothing is found about these diseases in the old writers. For this reason they have remained undescribed until now, and the cure too was omitted. Because man, however, is a discoverer of the origin of such diseases he can explain them by the light of nature; therefore a description of this disease is presented in the four following tractates. And although it is such, the lung sickness, as it may be called in German or in Latin, has not been described and explained correctly, as the light of nature in it proves. Even if they say that their proofs demonstrate what the disease in itself is, yet their proofs are considered in philosophy as something full of holes. Therefore know that the birth of the miners' sickness is like the origin of the lung sickness, as I indicate it. For the birth of the miners' disease is a testimony of the errors of the ancients in the writings on the lungs.

The second chapter.

On the birth and origin of the lung sickness

The matters are so to be understood, that you should know that the air is the body (corpus) from which the lung receives its sickness, and that outside of the physical air nothing harmful is added to the lung. Take an example, someone drinks and this drinking turns out to be harmful to the lungs. This injury does not arise from the drink, but from the air that is contained in this drink; it is drawn into the lungs and is consumed there. Every element has its own stomach in the body and in this stomach its element must be consumed. Thus the air is consumed in the lungs. And in the same manner as the stomach digests its food, one part being taken for the use of the body, while it excretes the other, thus it must also be understood of the air, of which the one part is also consumed, and the other part is excreted as an excrement. The air must be discussed, and it is entirely to be understood like a food, and as it is possible that food produces diseases, thus it is also possible for the air to give birth to these things. Therefore it is not necessary here

to describe anything else, but only to explain the chaos [1], since
it alone is the element that should be taken for comparison
here. In the same manner as you see that a chaos lies between
heaven and earth, which produces all the diseases of the lungs,
their fevers, ulcers, consumption, plethora, cough, gasping, and
oppression together with all the other kinds. Man must nourish
his lungs with this same chaos. Now the chaos is governed by
the power of the stars. It is therefore subject to their rule over
it, and as this species of ruled chaos is given to the air, thus does
it impress itself on the lungs.

This is now the basis from which the physician should recognize the diseases of the lungs, that they are there through
the power of the stars in the same manner as a disease which
is due to a food. There are also as many species in the chaos,
which make man sick, as many as grow in food; they are
also as well or as badly cooked as food. And just as it is
possible to prepare false dishes with food, thus is it also to be
understood of the chaos. As the example now proves, the
chaos lies between heaven and earth as a food for the lungs
in the same manner as the growths of the earth are food for
the stomach; that is why similar fruits grow above the earth
for those who live between heaven and earth. Now understand too, that there is also a chaos in the earth which rules
the lungs of those who live in the mines. And as those on the
earth become lung-sick through their chaos, so do those
become lung-sick, who are subject in the mines to the earthly
chaos. Thus the names differ according to the elements, namely,
lung-sickness for those who are on the earth, and mine-sick-
ness for those who are in the earth. The consequence of this
is the right to write another book. The upper heaven in its stars
is the one that cooks the air, which lies between it and the earth.
Likewise there are the mineralia of the earth the heaven and the
stars, and they rule their chaos in the same manner as the outer
heaven its chaos. And in the same manner as you see that
arsenic can kill us, thus there are stars which kill like it by intro-

[1] Gas.

ducing the chaos. These are the stars about which more must be written, how they produce the diseases of the lungs, insofar as it is possible for the air to poison its stomach.

The third chapter.

In order that you may understand the cause of asthma as quickly as possible, know then that the heaven is the element fire and that its elemental movement produces the chaos which shall be discussed here. And in the same manner as water is brought to boil by fire, the chaos is that which is boiled by the element of heaven. And just as the meat in water gives up its strength to the water, so the stars are like the meat and give their strength to the afore-mentioned chaos. And just as the soup from the meat is food for man, so the chaos, of which we are speaking here, is also food for man. Just as the food is digested in the stomach and has its special gullet, thus the chaos is digested in the lung and it also has its own gullet. Just as the things, that are placed in water, have their properties and make people sick or healthy according to them, so do the stars, that are placed in the chaos also produce a soup that is healthy or unhealthy. For this is the soup in which the plague is prepared, that enters through the lung tube and which proceeding further according to its anatomy, flies like a bird to its nest. If God had not determined at the beginning, that some should be protected and leave children behind whose seed should remain, who would have remained blessed then? Know then also concerning the lung-sickness that it comes through the power of the stars, in that their peculiar characters are boiled out, settling on the lungs in three different ways: in a mercurial manner like a sublimated smoke that coagulates, like a salt spirit which passes from resolution to coagulation, and thirdly, like a sulphur which is precipitated on the walls by roasting. In the same manner as you see a clean barrel, that is filled with clear wine in the fall and the wine has no palpable and congealed components. But at the end of the year, when the wine is poured out again, these three kinds mercury, sulphur and salt are found to have settled on this barrel; this is the winestone (tartarus). Thus in the same manner, just as there is something in the wine which was not seen in it, there is also a body (corpus) in the chaos, which attaches itself to the lungs, as to its barrel, and which then hardens there like a mucus in its viscosity, after which the coagulation starts, which is the matter of the lung sickness.

The fourth chapter.

I do not want to describe the species of the lung sickness, but what has been reported here is an instruction, so that you will understand concerning the heaven and the chaos and that they are also in the earth. That is to say that those who are instructed in the philosophy of the earth will also explain the lung sickness, since in its origin and rise the miners' sickness shares the species and the rise of the lung sickness, which is why the one explains the other and allows it to be understood. Anyone who has been instructed concerning the terrestrial diseases, also knows those in the firmament. One who writes correctly concerning the diseases of the firmament will also hit upon the diseases of the earth correctly. One who does not strike the earth will also shoot astray in the heavens. The physician should be so grounded in the light of nature, that he not only knows seven stars, but all the stars that the firmament contains. He doesn't stop with this knowledge, since he also knows the earth if he knows this, and therefore also the other two elements in their astronomy and philosophy. Therefore know further, just as the chaos is born, so is the earth a heaven of this generation, and the minerals that lie in the earth are the firmament of this heaven. Fire arises out of this element of the earth and makes a chaos in the earth of the same kind as the chaos between heaven and earth. And this same chaos becomes a soup of its minerals in the same manner as the external chaos is a soup of the stars. Now such people as seek and make their dwellings in the earth must carry on and nourish their lungs with the chaos that is there. And that which has been cooked in the chaos is the mineral impression, is the tartarus of the lung, which I call the miners' sickness here. And therefore the mode of origin (modus generandi) is the same process in both diseases, which ends according to the three kinds of mercury, sulphur, and salt, the type depending on the definite property contained in this flesh. It is well to consider this point, that this chaos acts in two kinds of bodies; since you know that the Earth's own inhabitants were made for her, just as we were made from Adam to live in the air between heaven and earth, just as you also know similar things about the nymphs. The chaos of the earth has been given to the inhabitants of the earth as air, the chaos of the water as air to the nymphs, they live therefore from this air. The one body is that of the inhabitants of the earth; for this I recommend the Archidoxa and the books Paramiris. But it should be understood, the other body is that of people who become inhabitants in the mines, and they are not people of the earth. From which it follows that the human chaos must be carried with them into the mines, since their lungs are maintained by their chaos, that is by the human chaos. But there it comes to a mixture of the earthly and the firmamental, and the two kinds become as one there, just as in a marriage; now the individual is suited for this constellation of the earth, from which constellation the miners' sickness takes its origin.

THE SECOND TRACTATE

ON THE ORIGIN AND BIRTH OF THE MINERS' SICKNESS

The first chapter.

The origin and the source of the miners' sickness are contained and described in the second tractate. First experience in these things should be made known, so that the things which give rise to the cough, the gasping, and the lung sickness with all its additions can be recognized and found through examination and clear discernment. The theory of the two sicknesses,

of the lung and of the mine, is divided according to the content
of this experience. In the same manner that you may comprehend the things which visibly demonstrate that they make the
lung sickness, thus they are also to be discovered according to
ascertained philosophy, to be present in the influence that was
described in the first tractate. You see that fogs grow externally
in the chaos between the heaven and the earth and these fogs act
in different ways, some of them producing asthma, coughing and
short-windedness. Now this is the experience which teaches us
to comprehend that the fog is the cause. Thus because the fog
has its origin in the firmament, there is also a fog in the mine,
from which the miners' sickness can arise even more severely
than from the external one. Now if the cause of this fog should be
sought, it can be found that it comes from the sphere Galaxae; [2]
those that make it are also in the earth. Now the mineral
of such a fog is also a cause, and the recognition of this mineral
gives the knowledge of the cure in the same way, just as a knowledge of fire tells us with what it can be extinguished. Thus
must all diseases be recognized, whose cure is then possible.
For this reason too death is incurable, since the heaven of this
constellation has never been found.

Therefore know further, just as it is to be understood concerning the fog, thus it is also with rain, frost and the like,
including also such a winterly cold, from which short-windedness can likewise arise. These things are all to be considered
in the mines.

The second chapter.

Now to speak further of the things, which make asthma, such
as cold and heat; for instance, a large heated lung, which is
cooled with sudden cold, is also attacked by short-windedness,
as also through sour beverages and through sweet ones. Now
just as such a sudden cooling of the lung, as well as sour and
sweet, make short-windedness, thus is this also to be understood in the mines, in that the work produces a heat in the
lungs, and the neighboring cold, which penetrates into the

[2] Milky way.

chaos, causes a rapid cooling of the lung after the work is finished. Although the cold is not felt, it is still essential in the alant and in the constitution [3] of the earth and its effect is such as if it had been drunk. The same is also to be understood concerning the acid. As you see that a sloe encloses its acid with a skin, thus the acid of the mine is also enclosed by a covering in the earth; and because we make dwellings in the mines we walk in this acid. Now the one acid comes from vitriol, the other from alum, as it can also be understood concerning the sloes and currants. The conjunction can arrange that the things, the acid and the like, are attracted into the chaos of the earth and the lung is eager for them; now it is injured in the same manner as one who has a special desire to eat chalk or another to drink vinegar and there are very many such desires. Thus just like these desires it attracts to itself, now the alum, now the vitriol, now the saltpeter, etc., and when the lung loses this desire, it fares like a sick person, whose desire turns out for the worst. Thus this acid causes hoarseness of the lung, like vinegar or a sour drink, and afterwards it is possible for short-windedness to arise. Just as you have now been instructed concerning the acid, thus should you also understand concerning the sweetness, which lies enclosed, like the sweetness in the currants; and as we walk in the mines, it is the same as if we ate this sweetness. When we eat the thing of the earth, the same thing happens, as if one were to eat currants with the teeth, the only difference being that the sweetness of the mine is ingested in the chaos. And when the desire misleads the lung, this sweetness produces the miners' sickness, the cause of which will be related in different places.

The third chapter.

Now the things are to be dealt with, which permit the external recognition of hoarseness, among which one is fat. If the lung has a desire for fat, it must expect harm from it, which then shows itself. Now there are several kinds of fat that we eat,

---

[3] Character of the earth.

from oil, meat or fish, visible or invisible. For this reason several lung sicknesses are found, since some become hoarse from this fat, others from another. Now that the eyes may see and comprehend, nature teaches that fat is also found in the chaos, under the sun, and also in the earth. Now every fat is nothing but sulphur that is divided in different forms and ways. From this it follows now, that the stars show their action in such matter in the same manner as the vapor from the intestine burns after being lit, and it is only an exhalation. Thus the chaos is also furnished with a fixed sulphur. If this sulphur is seized by the lung, it attaches itself to the latter like a resin externally to a tree. And according to the different types and kinds of minerals, different resins arise in the lung; this resin is the complaint and the cause of the miners' disease. Now the cause, why the chaos becomes a resin, which indeed is not its ultimate matter, is that the lung cannot digest it. In the same manner one recognizes, if the stomach digests poorly that it is weak; due to which various things can happen to it, and such things also happen to the lung here. In order that you may understand this sulphur in the chaos, take this example: you see the sheet-lightning during a thunder storm; now no sulphur is to be seen at the spot, but if it burns sulphur must be there. And it is possible for this lightning to blind, and if moisture were not innate in the chaos, it would burn down the houses. Now if this is possible and comprehensible, then such lightning can also come visibly, or also invisibly into the mines, and as it is possible for that which is visible to cause harm, then it is likewise possible that that which is invisible can also do so. Since just as a sudden exhalation which strikes into one can turn the lung to resin, thus it is also possible that such a vapor arises from the minerals, a thing which is often seen in the heavens.

The fourth chapter.

Now if this alone is the cause, that sulphur lies in the chaos, it is this that injures the lungs, as has been indicated; in the same chaos there also lies an admixed mercurial smoke, which gives the chaos its density and removes its transparency. It acts, as if a suddenly rising vapor of mercurial arsenic entered into one, for any such vapor produces a permanent, intense hoarseness; such a vapor also lies in the chaos and often, in addition to this one, still another is born. In the same manner sulphur lies in the chaos of the world, and to this the lightning is added. Thus two injuries are recognized. Understand also, that in the earth there is always a constant mercurius like arsenic as well as an accidental one due to the daily constellation. For the minerals are like the stars in the firmament. For this reason recognize what minerals the particular region possesses, since this is the influence; thus if Cachimia [4] were there, sulphur arises from cachimia. And just as if you wanted to say, the land lies under the Virgin, and just as the astronomer recognizes his stars and what the destiny of each is to be, thus does the earthly philosopher also recognize to what this mine inclines or which constellation is present. For the one earth bears its Marcasite, the other its antimony, and thus whatever the region contains produces the chaos of which we have already spoken. And just as the astronomer says this star exerts its influence, so too does the philosopher say there is orpiment in the earth from which this influence arises. And therefore the physician arises and speaks thus: This miners' sickness is a resin, which was born out of the chaos, and in which the star and the mineral orpiment are boiled. From this follows that the properties of all kinds of minerals should be recognized. For the sulphur of the marcasites is white and red, the sulphur of talc is yellow, red and black, the sulphur of cachimia is brown and black, the sulphur of pure antimony foam, the sulphur of marble, the sulphur of tufa, the sulphur of amethyst, and many others which it is not necessary to mention. And just as you have understood concerning the sulphur, thus you also have a mercury of copper, a mercury of lead, a mercury of ore which has already been roasted, a mercury of zinc, a mercury of arsenic and the like. And in the same manner as you have understood

[4] An imperfect metal.

6

concerning mercury and sulphur, it is also to be recognized concerning salt (sal). For there is Sal Vitrioli, Sal Aluminis, Sal Entali,[5] Sal salis Petrae, Sal salis communis, Sal salis Gemmae, and many others. This salt, this sulphur and this mercury are spiritus, which are the chaos itself, their nature accordingly corresponding to the particular region.

The fifth chapter.

Notice therefore concerning the spiritus, that the mercury yields a soot. Thus this spiritus also acts in the same manner as a sublimation which produces the ultimate matter. For this reason its miners' sickness is dry and lean and the spiritus sulphuris gives its resin and the spiritus salis its tartarus; these are the three chief types of the miner's sickness. Now although marcasite has one name, it is no less also cachimia, so that the former is not to be understood as being only of one kind. For the changes of the earth also change the minerals, as you see that the forms of countries change. Although an herb has the name cepa,[6] still it often has another form in another region and yet is still cepa. If now the form is different and changes somewhat, know then that the property in nature also changes due to the earthly power and also is differently changed in other parts of the same country. Therefore the physic of the same country is more useful than a foreign one. Thus like is healed by like, even if nature is vexatious. For just as much must be given to the minerals as to the heaven which rules the stars, thus also the earth and the minerals. And heaven and earth are two similar heavens, and the minerals and the stars are two similar stars. Therefore whichever is recognized in the one, is also recognized in the other. For this reason the physician begins with astronomy and ends with philosophy. For astronomy gives him the constellation of the firmament and of the minerals, and philosophy indicates the properties, both of which are necessary for the cure. Since man is a microcosm, he must be recognized through externals, and the

---

[5] The salt of solid saltpeter.

new moons, the exaltations, and the like are not only to be sought in the planets but also in the terrestrial stars. For if lunaria can demonstrate themselves in their course, then it is an indication that there are also more lunaria, just as there are also other things, which visibly demonstrate and present the heavenly course. Thus this tractate will now be ended, in which was described how the miners' sickness arises out of the three kinds of minerals, i. e. from their spiritibus, which are in the chaos that emanates from the minerals, like a voice from the mouth.

THE THIRD TRACTATE,

OF THE FIRST BOOK, CONCERNING THE SIGNS AND THE NATURE OF THE EXTERNAL AND INTERNAL BODY, AND HOW THE DISEASE ARISES

The first chapter.

Now in this tractate the work of the disease, how it arises, is to be dealt with; since man has a fragile life, and there are many things that kill him, from which he cannot protect himself even if he knows about them, therefore it is necessary to relate that which must happen through the physician, who has this at his command. For this reason God has created him, in order that he should recognize and explain the fatal kind, how nature strives against nature, how one thing in nature opposes another in the same manner as the animals on the earth band together against each other. For man should know this, that in the insensible things there is a uniform enmity. For the exhalations of such minerals also kill us in the same manner as the crocodile with his breath corrupts and kills man. However, because God has created good and evil, which cannot possibly exist next to each other and yet do exist united, he has set the physician in his place, in order that he may point out to mortal man the enemy who seeks his body and life, so that he will know how to protect himself. For just as the doctrine by means of which we know the devil, the enemy of the soul, comes from God and through the prophets and apostles, thus has he also indicated the enemy of the body, so that man may recognize by the light of nature what is poison, what evil, what good, and what is useful for him so that he may retain his long life. In order that the poison which is in the good should not break his life, he has given the physician knowledge to indicate good and evil in a thing, and subsequently he instituted Vulcanus,[7] so that good and evil can be separated from each other by means of this art. This art is like death, which separates the eternal and the mortal from each other, and this art should correctly be known as mors rerum; for that which should do nothing comes from that which should do something. Consequently it is right to describe the things that we may neither eat nor drink without worrying about the enemy, without worrying about the air around us, if we do not protect ourselves in winter and summer when we plant and raise our own poison in the garden. Besides we must also have gold and silver, also other metals, iron, tin, copper, lead and mercury. If we wish to have these, we must risk both body and life in a struggle with many enemies that oppose us. If we also want to have other things which we are forced to utilize for our healthy life, then there is nothing which doesn't bear our enemy within itself. Because so much lies in the knowledge of natural things which man himself cannot fathom, God has created the physician. Now out of the word arises the knowledge of the light of nature, and by it he can discover the good, and also recognize the good in the evil which is useful for healthy life. And in the same manner as the devil is driven out of man, the poisonous diseases are expelled by means of such physic, just as evil expels evil and good retains good.

The second chapter.

Therefore it is necessary to discuss the miners' diseases, and to present the things from which they arise, as was

---

[6] Allium cepa = onion.   [7] The art of alchemy.

indicated in the former tractate. Now our physic is in mercury, sulphur and salt, and our poison is also in these three things; for they both exist together. Thus we find through mortem rerum [8] that that which can help us in our distress, also has in itself that which created our distress. For example: Imagine a miner who looks for silver until he finds it, now he has an ore. Now it follows from this that because he seeks the ore and handles it he acquires his disease, which cannot be outside the earth, but only on the earth. Now if he has the disease and takes the same ore that he has hewn and allows the silver to be smelted from it, he finds in that which escapes the same thing that made him sick; thus he also finds by refining that which can make him healthy. Know therefore, that the vapour which escapes from the ore has the same poisons in it, as those which escape from the silver during smelting. For a rose can produce its odor and make us faint through its odor, while the body of the rose remains unharmed, therefore know the power of emanations when they enter into man. Although the body of this poison is not there, the agile poisonousness is there with such a weakness, that it drives man into a long sickness, which would be short if the body itself were there instead. For example: If arsenic is ingested there is a rapid, sudden death; if however the body is not taken, but its spiritus, the latter makes a year out of an hour, i. e. whatever the body accomplishes in ten hours, the spiritus does in ten years. Also there does not occur as terrible a death as when the body itself is present. Know that anyone who wants to acquire knowledge of the miners' diseases must know the rapid disease and the death that is produced by the body itself, with all the properties and signs that arise through the body. By these signs he can recognize what type of miners' sickness it is, i. e. the spiritus of which body is acting; thus the signs are recognized. For example: The taking of realgar produces a dried-up lung; by reason of this dryness the breath is changed to shortwindedness; together with this a blanching of the face

[8] The death of things.

appears. Fissures and crevices arise in the liver, and together with them an unnatural thirst which corrodes and grinds up the folds in the stomach, so that they peel off like the bark from a tree. At the same time a pressure arises in the epigastric region, and a difficult and hard digestion. This is followed by a high fever, palpitation and tremor in the region of the heart, subsequently an eruption on all the limbs, then the quinsy and an accompanying sickness of the head.

Therefore observe, what these effects of the poison are and with which disease they appear. Thus over many years and days the spiritus of realgar also produces such a uniform disease, which depends on the previously reported accidents. Such infected miners tend easily, according to the courses of the external heaven, to all heavenly diseases, such as fever, consumption, to raving madness and the quinsy.

The third chapter.

Thus after the effect of the poison has been reported, how it causes the disease, it is necessary in this tractate to recognize the signs that come from the bodies, as it is written above; there are so many of them that their number cannot be counted. But notice briefly, that under realgar, are understood and comprehended all the arsenicalia, including also the operimentiva.[9] Anyone, however, who can easily fathom the separate species because he has had much experience for a long time in these diseases, can distinguish and divide the spiritus with greater certainty. The treatment will be discussed in the chapter on realgar in which the arcana will be explained. If, however, the cure should be detailed, it is necessary to specify these exactly, as the humoralists do. Thus just as there are various species of realgar, there are also species of antimony, to which marcasite, cachimia, talc, pure antimony foam and the like belong. For if the poison antimony is taken, it produces a dry, barren cough, sharp stitches in the side, headache, constipation, an ulcer of the spleen, heated blood, scabs and itching, drying

---

[9] Yellow sulphide of arsenic.

up and increasing jaundice. As far as the signs of the above mentioned minerals are concerned they are to be ascribed to the spirit of antimony. You should also know that under the name alkali many minerals drive their spiritus into man in a poisonous manner; thus in the chapter on alkali the signs of this death are given: it causes difficulty in breathing, foul odor out of the mouth, and throws out much mucus, produces heartburn as if it were in the stomach, causes diarrhea, griping in the stomach, and aching in the belly, dries up, causes the lungs and the stomach to putrefy, splits the liver and the spleen, melts the kidneys and makes the urine sharp; through its devouring corrosion it causes the region of the kidneys to putrefy, expels pollutions and also blood through the urine, and where diseases lie in such places and limbs, they are moved and driven outward. The species of the blue and the white vitriol belong to the alkali, and the three species aluminis rochi,[10] scissi,[11] plumosi,[12] also the species salis communis, gemmae, silicis and whatever else belongs with them. Now observe, after this instruction, the differences between these ores, according to their species. For there is a realgar of gold, of silver, of iron, of copper, of tin, an antimony of gold, silver, iron, etc. and also an alkali with the same differences. It is necessary to know these differences because of the cure, and also the type of the country should be considered. For the eyes demonstrate great differences in these simplicibus, for instance the Hungarian region and the Styrian region are different in their characteristics, but ultimately have the same metals. Thus too the Etsch mountains and the Inischen mountains are in a different class than the high mountain of Meissen[13] and the same is also to be understood of others. And they differ even more than Rauris and Gastein, than Pinzgau and Pongau, and even more than two galleries side by side. Such things are recommended to experience, without which these diseases cannot be understood. And it is useful to have seen all the corpora

[10] Mica.   [11] Crystallized alum.   [12] Talcum.
[13] Mysnia = Meissen; Etsch and Inischen Mountains are in Tyrol.

as they are found in their species, to have observed their action, and how they produce their evil and poison. For the marcasitic sulphur prevents sleep, the vitriolic sulphur produces sleep unto death. Thus there are also vexatious salia and also mercurialia; this I recommend further for the experience of each one.

The fourth chapter.

And although there are many enemies so that the physician cannot indicate their number, yet it is not necessary; for anyone who would want to compute such things would introduce too many errors into his practice. For the practice is such that it fights against the disease with the Arcanum. Observe therefore how you are to understand the arcanum: you know that there are various arsenicalia and the like which arise from gold; now whoever wants to make a special prescription for each of these species, departs from the arcanum and produces serious errors, but he who attempts such and considers the gold, approaches the arcanum; for good heals evil, which stands next to it. For instance: whatever causes jaundice, also cures jaundice.[14] It is thus: good and evil are in the same thing, the jaundice arises from the evil, and when the good is separated from the evil, the arcanum against jaundice is there. And therefore it receives the name arcanum, because out of the same evil many kinds of jaundice can arise. Now whoever wants to consider each jaundice separately does not know what the arcanum is; for the arcanum cures all these species. If the specialist (particularis medicus) wants to seek a special diet and a special prescription for each one, he will seek too long to be able to help the patient. Thus according to the type of this cure and order is the origin laudani, i. e. materiae perlatae, which is the same as if a pearl is extracted and this pearl cures the disease which arose from its evil. For the physic which should cure paralysis must come from the same thing that caused the disease. Now if one acts according to the arcanum, the worry concerning the many species of the same disease is removed. Thus the arcana are

[14] Homeopathic principle.

also comprehended here in the minerals, e. g. gold is also a physic for all the diseases that its miners acquire. Thus Saturnus also has his arcanum for the diseases which arise from lead within him, and all else is also to be understood in the same manner; what can be harmful through our hands, can also be made into physic by our hands. And whatever is impossible for these arcana will further be found to be impossible. In the same manner as we see that stab wounds and other wounds can bring about death, whereby no advice and no help can be found, so too the minerals penetrate the limbs of life, so that they can be compared with such deadly wounds and stab wounds. Thus I now wish to close this tractate in which that with which we dealt has been explained sufficiently. For this reason, now that asthma has been sufficiently recognized, how it arises between heaven and earth, followed by the origin of the miners' disease in the earth and how far such poisons act confusedly, these things must be left, and the cure of such diseases must be explained.

THE FOURTH TRACTATE,

CONCERNING THE CURE OF THE MINERS' SICKNESS AND ITS NECESSARY PARTS

The first chapter.

Now in order to consider the welfare of the miners so that they can be protected from the aforementioned miners' sickness it is no longer meet to speak with the learned men and the philosophers, but with experienced men; for it is the manner and the innate custom of any experienced man not to confront another experienced man with idle talk. The health of the patient is considered in few and short words by the experienced, since the words of idle talk cannot cure. They cannot please the patients either, nor can the latter love these words. Experience is so constituted, that an understanding of its works makes itself known to everyone without much gab. For this reason, more attention should be paid to practice to determine what it is. It should be left alone, and this experience should defend itself and the results which should move every unbeliever to believe in physic should be examined. For the results are so clear, that they are not in need of any disputation; what these are, will be considered later. However, each one should retain his own experience; for who can or wants to fathom the end of medicine? For the school of medicine exists and is constituted without an end.

The second chapter.

Now I will first deal with the articles of the miners' diseases divided into chapters. First the mine spirits must be forestalled so that they will leave the miner uninfected; this is begun before they become subject to the mine and the ore spirits. Secondly it is good to know that the organs of man, the lungs, the liver, the kidneys, the stomach, the viscera etc., as many as the body contains without exception, are poisoned by the ore spirits in the course of time; they devour the chief organs in the body whereby these begin to putrefy. Now this putrefaction cannot be cured by replacing the lost part; but how it may be prevented without any further harm, like an estiomenical wolf [15] injury, is presented with its preventive measures below. Thirdly there follows, since it belongs to prophylaxis, a separate chapter on diet and its arrangement, by which man can maintain himself so that he does not through any cause make himself subject to the ore spirits. Fourthly, the natural cure with medicaments and powers, which dispels the natural asthma, is to be described, since this is also the cure of the miners' sickness, for which the prescriptions are mentioned below. Fifthly there is the cure with the arcana; as it has been mentioned how everything good has three kinds of marcasites, in order to dispel the three kinds of evil. Finally the close of the first book.

[15] Devouring wolf.

The third chapter.

On the preservative of youth

Understand this by means of an example: The heaven is that which infects us by means of its astrum, to the extent to which it has been assigned to harm us. Now we cannot withstand it, for it is not our power, that is we cannot forestall the influence. Just as we cannot prevent the rain or the snow, we must also let the action of the ore spirits proceed. But observe this: Just as a roof is found in the rain and a shade in the sun, thus in the same manner a chimney will be found through which the effect cannot penetrate. Understand this therefore: Everyone, who is still free from infection, is maintained thus by means of the Essentia Tartari, taken and used once every month, together with a good sweat; the prescription is as follows

> Rec. Liquoris tartari ℥ ij.
> Olei Colcotarini Ə j.
> Laudani purissimi ℨ fs.

to be mixed and administered up to the weight of three barley corns.[16] It can also be that according to the health of the complexes once in a half year is sufficient, but this is left to the discretion of the experienced physician. Moreover Perlatum Auri is better than this and all else.

The fourth chapter.

On the other preservative against putrefaction

For those who have been attacked by putrefaction, as described above, Manna calabrina perlata is best. Now Manna is every sweet substance that is extracted from each thing; this manna is of the species Balsami, for this reason the type and the power which do not allow putrefaction are present. By means of this Manna the body must be protected from putrefaction and it must resist alone the destructive putrefaction

[16] Used as a small weight.

of the miners' sickness. Now it is further necessary to discover this Manna, and one is in vitriol, one in urtica [17] also one in magnete. These three manna are the balsams which prevent putrefaction here, as is indicated. Now their extraction is as follows: the manna vitrioli is extracted so long by means of its phlegma until it acquires the sweetness of honey and a brown color. The dosage is one drop in aqua veronicae every day. The description of manna urtica is as follows; it is alkalized, then rectified so long by means of ventrem equinum until it acquires an oily consistency, at which time it is supposed to be free of all dross. Then four drops should be taken every day as indicated above. The extraction of manna magnetis is as follows, it is placed in alcohol, covered with an equal weight of limatura chalybis [18] and then glued together with lutum sanguinis [19] in a glass vessel and boiled in the oil of iron dross for one month. Then the limatur is taken from the magnets and this same limatur is digested with rectified wine until the wine becomes red. Then the wine should be separated from the red substance and the manna magnetis remains on the bottom; it does not allow any important organ to putrefy and is used as has been indicated above concerning vitriol.

The fifth chapter.

On the nature of the diet

Although it has been the usage of physicians until now to prescribe food and drink, it has none the less been in diseases which they understood, and such will be conceded to them and left out of the discussion. However, here in the miners' sickness it cannot be introduced, since its cause is another heaven, not the external upper heaven, but the heaven which is the earth itself. Therefore know that each heaven must prescribe its food in its realm, therefore that which is on the earth remains for its upper heaven. But that which the terrestrial heaven rules must taken its diet, that is food and drink, from

---

[17] Nettle.   [18] Iron filings.   [19] Blood lime.

it. However, because man is not born for this nourishment, but for that on the earth, he retains the external regimen and protects himself from the external asthmata so that the internal one should not be added to the external one. Therefore he should protect himself from the things that cause asthma, as has been related in the previous tractates, and observe also in this connection, that heat cooled in the mine with the same water does not produce an external lung sickness, but rather miners' sickness; the same is also to be understood concerning the other things. The diet, however, which corresponds to the type of mine, should be kept together with the others, and they should take great care, if possible, to salt their food with the salt which comes from the saltpeter that grows in the mines and which is boiled and prepared like ordinary salt. A salt should also be boiled from alum, to be used in place of spices; for it does not salt, but spices. With this I have indicated the salt and the spice that lie in the ores; I leave it to the experienced man to improve the things and to add more than I report and indicate.

The sixth chapter.

Now the nature of things should be held in high regard and since things must occur in nature, it is necessary that not only the authors with their glosses be accepted as trustworthy, but that nature be investigated further, for from it comes the correct teaching and the correct instruction. Thus one finds in her that in asthma only the arcana should be considered and not the composita. For this reason, because the arcana are devoted to man, they are also to be used in the miners' sickness. Although ruled by another heaven it is still important that man is of the upper world and can defend himself by means of such arcana, so that the ore spirits cannot withstand him, when the natural arcanum is brought into the mineral species. And although the mineralia have their own physic just as the surface of the earth also has its own, still man depends on both since he subjects himself to the dwellings of both. Now aqua panis porcini [20] is a

[20] Water made of Cyclamen roots.

special arcanum which cures every asthma that does not putrefy.
When the lungs are brought into a diaphoretic [21] condition,
nothing evil is attached to them. Thus too with the liver, the
stomach and other members. For if they can be maintained in
a diaphoretic condition, no ore spirit can harm them any more,
and the damage that it did will be expelled by means of the
diaphoretic nature. For to that which sweats no mucilage
attaches itself, no resina nor any tartarus. But if the sweat
producing power is lacking, the three reported things that can
cause harm spread not only to one member but to all. For by
means of the vesica diaphoretica and regio diaphoretica neither
gravel nor stones are deposited. This diaphoretic species, which
is treated here very briefly, will be dealt with in other places
at greater length; for those who have a knowledge of nature,
but not of practice, the treatment has been sufficiently indicated.
Whatever else is to be understood in such a manner in the ore
world follows herewith.

The seventh chapter.

Now it is necessary to speak further concerning the matters
of the internal earth which should also be understood thus,
that the main thing is to bring the injured members the power
to sweat, so that they will sweat like a skin, which opens in the
bath. And it would be the same if they were covered with winestone like a barrel or with viscous pitch or with alkali, nothing
would remain on the skin if a sufficient sweat would pass out from
inside. For this reason only the diaphoretic nature of the members that are covered with these three kinds, realgar, antimony
and alkali, is considered here. Realgar is the soot, antimony
is the mucus, or in order that it be better understood, the var-
nish, and alkali is the tartarus: these are the things which dis-
turb the air-containing organs. In addition, observe that three
kinds of arcana perlata can be extracted from the metals, and
each retains its diaphoretic property, which belongs alone to the
aforementioned species, so that the air-containing organs can

[21] Sweating condition.

be brought into their own natural water, by means of which water the three aforementioned things are washed off and out through the excretory openings.[22] Now because the ore dispensary has no species besides its metals the cure is thus, that the vitriol is extracted from the gold and the sulphur by means of the (drotten?), and its prima materia should acquire the appearance of argentum vivum liquefactum, a brown color, and should dissolve in the fluid by itself without being moistened. And further, just as you know this about gold, so you can also understand with regard to the other seven metals, but the preparation is different and is sufficiently explained in the book of preparation where de morte rerum is treated, so that it is not necessary to relate it again here.

The eighth chapter.

And if the first tractates will be difficult and strange for the ordinary physicians, the cause is that the mines and that which belongs to them are also foreign to them, therefore it is reasonable that I proceed with experience in the light of nature. Since every doctrine comes from God and physic is created for the patient, it should not be concealed nor remain hidden. Therefore I have acted fairly and properly in the first three tractates, but concerning that which I have treated in the fourth it can be that some will complain, for the physic is not prescribed according to the usual order; to them I give the answer, if the usual order were useful it would not have been forgotten. However, because the disease remains incomprehensible, it is in need of this arrangement of mine; for the results make it trustworthy. But I do not doubt that the prescriptions will be difficult and very difficult for some to understand. These should remain satisfied with the answer, which is thus: for those, who are associated with the mines and in the mines, especially for the masters of the mines and for those who are experienced with metals, enough has been said and they understand it sufficiently. For if those understand a thing, whom it con-

[22] Emunctoria.

cerns, I consider it sufficient. For how can a silk-embroiderer turn a rope-maker with his cords into a silk-embroiderer? Therefore everyone should be shown that which concerns him. Thus the disease remains for the mine, and the book too for the mine, therefore the understanding also is to be acquired in the mine.

The end of the first book of the miners' sickness.

THE SECOND BOOK OF THE MINERS' DISEASES

concerning the smelters, refiners and silver refiners and others, who work with metallic fire.

THE FIRST TRACTATE

ON THE MATTER, WHICH HARMS SUCH PEOPLE

The first chapter.

Each thing, which is thrown into the fire has two kinds of components in it, a fixed one and another which is transient. Now we don't have to talk about the fixed body here for nothing harmful arises from the fixed bodies, but they remain harmless for man during the work in the fire. Further, however, this book will deal with the transient body, which is always together with the fixed one; this transient body is only found and recognized by means of the fire. For in the fire the separation of the fixed and the transient bodies takes place. Now because man separates the things, the things that happen to him while he does so are to be undestood as a result of this action. For the transient ones are not without poison and evil. Each good thing which should be attained must be separated from the evil. Now it is the usual thing that there is no love without suffering, for so strongly does the enemy maintain himself once he has penetrated into the good. Whoever wants to have the good must also expect the evil. Now man seeks metal so much without considering the injury to his body, and strives so intensely for these things that he ventures among his enemies and sees that he is surrounded by them, but nevertheless he remains because of the good which lies together with the poison. For one who smelts see the fumes from the ore, that they are a poison, and he sees that they are arsenical fumes and he smells that there is no good in them. Yet despite all that, he forgets his health, how inimical the poison is to his body, and he doesn't consider that his mouth and nose are open, and that the breath enters into them and with the breath these fumes. Only after a long time does he see and feel that the poison will be his death. However, despite that, God wants that the treasures and wonders which he has demonstrated in the metals be investigated and discovered, for which reason he has indicated the art of discovery. And he has also taught the necessity of separating the ores of silver and of gold. Now since these things have been arranged by God and since the diseases arise nevertheless, then it follows that the physician has been created because of these things, so that he may prevent, forestall and investigate the diseases that arise according to God's plan. For God is so good that he never forsakes us; only if we fathom his mercy do we find great virtues in physic, that these virtues overcome all poisons. Upon this godly mercy of the great immeasurable virtues of physic, this book is to be arranged and several tractates are to be written in which the nature of the disease together with the properties and nature of the transient body are contained, followed by the cure and the physic which rules in such diseases.

The second chapter.

This is the nature of the transient body; it has three bodies within itself, namely the sal, sulphur and mercury. The three are separated from each other in the fire. And although the fixed body is also present in these three, still the fixed is separated from the non-fixed. Now when the transient body prepares to separate due to the power of the fire, one part goes into the sal; that is the ash and the slag which arises there. I do not intend to write about this.

Further, the body sulphuris and the body mercurii, these two are the topic of this book, for the body sulphuris is the fire, since nothing else burns but sulphur; the body mercurii is the smoke which arises from the fire; the diseases of the people who work with fire arise from these two bodies. Now it must first be considered what is the body which is thrown into the fire, since the species of the fire and of the smoke arise from it. For each metal has its own smoke and fire. The smoke of ore which has already been roasted is different, just as the smokes of copper ore and lead ore are different, etc., and although only sulphur burns and only mercurius smokes, yet there are indeed various species, just as there are various ashes and slags; experience teaches one to recognize these things if one has witnessed them. For this reason it is first necessary to deal with the form of that which burns and smokes. Each fire has its air, which is innate in it and which arises from the sulphur. This air has its definite nature, like an air in which there is an impression. This impression arises from the sulphuric ore in the same manner as a lily gives off an odor which is innate in it. And just as this odor is a natural impression which does not separate from the body, thus there is also such an odor in the sulphur, which odor is called air. From this it follows that those who work with fire take in the air which arises from the fire and not the air of the world, in the same manner as when a fog is present, which is also a special air. Therefore know that the transient corpora make a peculiar air in the element fire, with which one can also maintain oneself, just as with the ordinary air which we receive. This is proven by the salamander, which does not maintain itself with the air by means of which man lives, but by the air which is peculiar to the fire. In the power of the element the salamander has its breath, and outside of the fire it has no life. Therefore understand also, that a peculiar air element is born from the sulphuric ore. And although this air is only in the fire and not outside of it, yet when the fire gives off warmth it also gives off the property of this air like the lily its odor. What can be said about the odor of the lily is also to be understood concerning the admixed element air.

The third chapter.

Thus you should also understand concerning the body mercurii, which does not remain in the sal, nor in the body sulphuris, but which flees the fire. Know further that if it wants to separate, it becomes a smoke; this occurs as a result of the separation which takes place due to the fire. Now it is its character to be changed so by the fire, that it becomes unknown to many. For this reason, however, it is discussed here. This smoke attaches itself, from which it may be recognized that it does not disperse or vanish. Although the fire loses it, also the sulphur and the sal, yet it seeks the cold and deposits itself; during this process one discovers of what species it is, for every smoke is a mercurius. At the same time that it is now recognized and discovered that the smoke does not disappear, but remains, it is also recognized how poisonous it is. Thus the smelting smoke occurs in many forms. What arises from the ore is to be recognized as arsenic, as realgar, as operimentum and the like with many other species. Now if we consider that this smoke is a mercurius, and that this mercurius is arsenic, realgar and orpiment, and if we know what dangerous poisons the three are, how dangerous they are for our lives, then we can know and recognize that contact with them is unhealthy; therefore it is proper to consider that the smoke contains the same poisonous powers before it is deposited as well as after it is deposited. And we see obviously that the air and the smoke become one thing and change into an inseparable mixture. And since man must have the air, so he must also have this mercury. For it is impossible for man to separate the air, but he must breathe it in as it is mixed.

The fourth chapter.

Now the question must also be dealt with: since it is not possible for man to separate the air, then that which the fire drives into air remains in the air, for the air does not consume the smoke. Now since the mercurius remains, it is attracted by the upper stars from the earth up into their regions, in the same manner as the needles of compasses are attracted in an opposite manner by a southerly magnet. Therefore know now, that the stars have in themselves the fleeing mercurius, which has been driven away by the fire, and which they too cooked and prepared before it existed. This they prepare for the second time and change it in the constellation and impression; this union is a mother of the febrile diseases. But since it is not intended to explain them here, but only what concerns those who work with fire, I recommend for this the corresponding books dealing with the subject. Now further, since the smoke and the air furnish the breath, the evil must now be considered which arises from such an inhaled breath, which is naught but mercurius peccans. Therefore notice the dichotomy which occurs here. The right air carries out its right office without being hindered by the mercury. In the same manner water which has been adulterated and is drunk against thirst, accomplishes its office and quenches the thirst without the aid of the adulterated thing, and thus it is also to be understood concerning this air which separates from the evil that is in man, in his lungs. This separation effects several ultimate matters, i. e. the mercurius enters and the walls and the region of the lungs dry up in time, and after this drying up which it causes, it precipitates itself upon them. And since the lung together with its region is comparable to a venter equinus [1] with all its properties and powers, for this reason the mercurius must undergo a change, and change differently, due to the power of ventris equini, than it does in the smelting works. Thus compare it with a food which is excellent and delightful; as soon as it enters the stomach, however, it experiences a dreadful change, which is effected by the venter equinus stomachi. And when it passes further into the intestine it experiences another change, this is carried out by the venter equinus intestinorum. Notice therefore that such various digestions occur in the body, and one is also in the lungs.

[1] A vessel for distillation.

The fifth chapter.

Now it is proper to speak thus. When the mercurius enters
the lungs, it chooses a wall to which it adheres, just as the
case or the situation happens to be. Therefore concerning the
attachment there is not much to explain as far as the place is
concerned; for in anatomy many sublimations are found and it
is useful for each one to know them. When it enters the action
of the venter equinus starts. Some mercurii are changed to oil or
fat, some to a jelly-like matter, some to a tragantic solution,[2]
some to a foenugrecian mucus, some to a leather-like glue, some
to a horn glue, and there are many such kinds. Therefore it
is necessary to know where such putrefactions arise, so that one
can recognize into what ultimate matter the venter equinus
pulmonis has converted them. For something, which enters a
region and is not the food of this region, cannot be digested.
For this reason putrefaction ventris equini follows, whether it
be thin or thick or in whatever way it can happen. Therefore
it is necessary to consider whether this putrefied solution is
deposited in the tubes of the lungs or in the membranes, and
how this solution comes there. And sometimes it occurs that
such solutions due to their acrid putrefaction begin to penetrate further, by means of which penetration such fluids pass
through the pores towards the region of the kidneys. Due to
this penetration this ultimate matter is found in the urine,
sometimes oily, sometimes in tragantic solution, sometimes like
a psillian [3] mucus, and they are often considered kidney diseases.
It also happens that such matters are also found in the sweats.
These things are all found according to the subtle putrefaction
of the venter equinus in which there is a great artist.

[2] Tragacanth or milk-vetch.
[3] Flea-bane.

THE SECOND TRACTATE

OF THE SECOND BOOK ON THE METALLIC SMOKE

The first chapter.

It is furthermore proper to finish the things that have been started, inasmuch as we have already spoken of the transient body to which is due the separation into good and evil. The evil is so firm in the good and united with it that they cannot separate from each other. For the metallic smoke, as well as its fire, and also its slag, are alone the three transient bodies that have been mentioned, but they are half-fixed because of the fixed body with which they are united. Thus there is among the metals a half-fixed salt, a half-fixed sulphur, and a half-fixed mercurius. From these three it follows that their fixed character is withdrawn and taken from the metals, and by means of the three every metal is consumed. For these three are the death of metals, since that which is attached yields slag, that which glows is the sulphur and that which fumes is the mercurius. For this reason then that they are so closely related to fixation, they have a metallic consumption, i. e. a metallic fixation which does not proceed any further, since the three transient bodies consume the metal in the course of time by means of the fire. Also gold and silver do not act like other metals, for they do not leave any slag nor do they leave any rust, Therefore know the difference by this: although the eyes do not see the sal of gold and of silver, still the departure is noticed by means of the weight which is visible in copper and iron since it disappears coarsely, but which steals away subtly from gold and silver and at the very last causes it to be consumed because of the sulphuric body which consumes itself in the fire. For the flowers and the pretty colored flames (flammulae), that accompany the refining arise from the sulphur which is in the gold and the silver. These same flames are the consumption of the metal just as such fire colors also appear in the case of iron and other metals according to the species of their sulphurs. Thus it should also be understood concerning the body mercurii, that it disappears in the manner of smoke. For every thing which lies in the fire has a smoke corresponding to the metallic fixation. Therefore know that the metals emit such sulphuric and mercurial spiritus, while they are being wrought and used in the fire, but they are so subtle that nothing is noticed without special attention.

The second chapter.

Now since there is thus a fragility in the fixed metals which is not as transient as the first one, know then that those who work with the metals are capable of receiving the same. Notice further, that they have a more subtle, sharper passage than those first mentioned. For this reason it should be noticed further that they cannot be recognized like the smelting fumes, because of their subtility in the lungs. Because these spiritus are so closely related to the fixation, the result is that the venter equinus cannot do anything to them. For this reason their ultimate matter is such that they penetrate the lungs and these parts in the same manner as a tincture (color) which consumes itself in the same body, so that they become *one* thing, the body and the tincture, and the body assumes the property of the tincture. In the same manner as when copper assumes the color of zinc, for which reason it is then ruled by the property of the zinc, yet the copper remains copper for the zinc cannot consume it. However, when the lung is stained it cannot withstand the tincture, but the tincture eats it and gnaws it away, just as an aqua fortis which is poured over a linen cloth destroys the latter, and thus it also happens to the others when they receive the tincture, especially the neighboring region of the lungs and the stomach. At the same time this metallic spiritus passes through the entire body and where it attaches itself it acts upon that place, as is indicated above. For this reason various diseases are met here that cannot be described otherwise than as has been said above; they are to be recognized by their devouring character and by the many strange signs, that cannot be described with certainty because of their variability, for their species vary so much, that only great and long experience teaches and indicates them. However, every physician should know that the diseases of such people are all related to the metallic kind. For the tincture of these two, sulphuris and mercurii, is so bony and pointed, that it consumes all natural humors, qualities, complexions, and colors them all with its character. Therefore the physician must recognize the tincture and not the humors. For although it happens that all kinds of diseases accompany it, such as dropsy, jaundice, water in the joints, fever, etc., still they are all ruled by the metallic tincture against their usual innate character. Therefore it is so necessary that the physician have more experience in these things than the ordinary course indicates. Metallic diseases are not humoral diseases.

The third chapter.

Now these spiritus cannot be separated by man, but because he is in the air, he is also in these spiritibus, therefore know further, that the air not only enters through the lung tube but also through the nose. Now the most subtile part of the air, which is inspired through the nose, is not precipitated in the pharynx, but rises into the brain and its regions. For you should know this, that the brain must have air as well as the lung; and just as it is understood and has been explained concerning the lung, thus you should also understand it concerning the brain, that in the brain too putrefactions, tartara, realgar, fuligo etc. grow and arise. Therefore the nose exists to furnish the tube through which the air goes into the brain; the most subtile and the coarsest inhalations precipitate themselves in the lungs, but when the nose is clogged and doesn't allow any breath to pass through, they pass upward from the pharynx, through the internal nasal orifices. Therefore we can understand the headache, the head ulcer, the head sickness and other diseases of the head that arise in this way, such as paralysis, lethargia, tortura, etc. Now although these diseases exhibit and reveal themselves according to the old nature, they are still like venus album [4] or aurichalcum,[5] that is, if the tincture were not there, it would be the old disease; but the tincture, that cannot be separated from it without harm, gives the name here, so that such a paralysis is called paralysis mineralis. Thus likewise lethargia mineralis etc., and the same with the others. For just as various diseases of the old nature are found, thus various tinctures can also occur. Thus too it can not be contradicted that the same is likewise to be understood concerning the first smelting fumes and also concerning the brain; because in these fumes there are also many subtile spiritus that can be mixed with the subtile air, which is determined only for the brain. Then come brancha, pituita, coryza, catarrhus and others. And thus as it has been explained concerning the brain and the lungs, you should also understand such an action in the stomach. And although various things could be contradicted, still the anatomy of the stomach reports it thus, visibly showing fuligo, tartarus, realgar, as well as bitumina, muscilagines, galredas and others. However, such a hidden descending air enters into the stomach, that one can hardly conceive of it or perceive it without proof.

The fourth chapter.

Now in order to explain the things further from one handicraft to another, know then that every handicraft has its metallic spirit, thus, those who work with iron succumb to the spirit of iron and those who work in fire with copper succumb to the copper spirit. Thus too with the other metals. And those who work with brass succumb to the spirits of copper and zinc, just as those who work with albo [6] and rubeo [7] to this metallic and colored spirit. Now know that these spirits are known, so that it is not necessary to speak any further about them here. But some experience with the handicrafts that work with metals is necessary. Thus those who refine by cupellation while refining silver receive lead fumes and silver fumes with

[4] White copper.   [5] Brass.   [6] Silver.   [7] Gold.

the properties contained in them; the silver refiners receive only silver fumes, those who make minium lead fumes, and those who make white lead and lead ashes the acrid lead fumes. For when a metal is about to break, it drives its strongest spirit from it. And those who pour antimony succumb to a hard sulphuric and mercurial smoke. Similarly those who pour zinc succumb to various subtile fumes: to those of Venus, Mars, Jove and to that of an immature ore, which has already been roasted. And those who make ultramarine succumb to the most acrid silver spirit, which is mingled with an admixture; those who work with litharge succumb most seriously to the lung sickness; those who work with cinnabar succumb to the spirit argenti vivi and sulphuris.

Therefore understand further concerning the metallic handicrafts how they poison and attack their masters and vulcanic slaves, and with what a variety of species and ways; the more the art is developed, the more enemies and accidents there are. Thus we see that when we do not use fire subtly, it does not poison us subtly, but if we seek very subtly in it, the fire opposes very subtile poisons. For example, the stellion crawls into its hole and harms no one; but if we wish to seek in it, we must expect much opposition, so much so that life and limb must be risked before obtaining from it that which makes one rich. This also happens to all those who divert and enjoy themselves with fire, so that finally everything turns to grief and sorrow. For these things are like the Epicureans who enjoy themselves with good food and drink; they must pay for it with serious diseases. And anyone who is happy with his pretty speeches will also be insulted by that. Thus no love remains without hurt.

The fifth chapter.

Therefore observe, as has been said, that we have three regions in the body in which these things divert themselves, the brain, the lungs and the stomach. Know then that they do not retain their evil in themselves, but send it further into the body, for which reason all the more diligence must be applied to understand the mineral diseases. For as you see, if the brain suffers from a mineral disease, the entire body also suffers. For the character of the spiritus is such that they gnaw in the same manner as a worm in a finger. Such is the mercurial character that wherever it centers itself, there it gnaws and tears like a burning coal. This tearing and gnawing is a cause of many diseases, such as mania, phrenesis etc. Thus it also occurs that an orexis [8] arises in the lungs which is like a panaritium, and in the stomach too it gnaws and tears continually around the pit of the stomach, as if a biting worm were there. And not only does it cause this gnawing and tearing, but also many more effects, whose species and names have never been recognized or mentioned, and which cannot be named here because of lack of room. But everyone should endeavor to be well instructed concerning this disease according to the old nature, consequently one should gather experience in these mineral diseases and learn in those places where these diseases are and live; for the physician becomes learned through this exercise and experience. For even if I would report and describe everything, still no one would be able to understand it without the experience. Thus if he wants to have the experience, let him acquire it where it is, that is among the mineral diseases. For who could be taught the knowledge of experience from paper, since paper has the property to produce lazy and sleepy people, who are haughty and learn to persuade themselves and to fly without wings, all of which things are repugnant to the physician. Therefore the most fundamental thing is to hasten to experience.

[8] Cramp or gnawing pain.

THE THIRD TRACTATE
ON THE DISEASES OF THE SALT ORES, NATURAL AND
ARTIFICIAL

The first chapter.

Since all the ore diseases of the mines are supposed to be included in this project, the salt ores ought also to be explained among them. Now they do not act deeply in the body, because vapors that dissolve in the warmth are the cause of their spirit. And even if they produce a moist air, as happens, and come to the three places mentioned, to the brain, lungs and stomach, together with the other air, they are harmless and more likely to advance health than to create diseases. For the causes of the salt species are of three kinds under which all other kinds are included and named, namely salt, vitriol and alum.

In order that they be tested to show that they are rather healthy than unhealthy internally, note this. When salt itself passes through the nose, it produces sneezing; if not, it penetrates to the brain and dissolves the phlegmata, the mucus and the apostemata, so that these cannot collect in masses, and dries the head internally, so that a good, healthy and dry head remains. If it then strikes the lungs, it dissolves the same things in them, the things which produce coughing, shortwindedness and fullness. Also if the lungs are about to ulcerate, it prevents it and is like a balsam at this place; it doesn't allow anything to putrefy or to be deposited; whatever wants to coagulate and to become inspissated, this moist air dissolves it. Thus too when the salt in this air enters the stomach it cleans it of its moisture, although this is otherwise not the character of this salt when it is taken alone, but because it is mixed with the air, the highest essentia that are contained in the salt can act. For the most subtile essentia are those that become an air. Thus the salt is also useful to the stomach in digestion; it protects the latter from mucus and confused masses, and maintains a good appetite; besides that it is also beneficial for the eyes, ears and teeth.

Thus you should understand the same concerning vitriol, that when its spiritus enters the air by means of its dry fumes or through boiling, a moist air also arises from it. This air is the most subtile essentia vitrioli and has the same properties as salt in the brain, lungs and stomach. Besides this, however, there are secret arcana in it against many serious and major diseases, such as jaundice, overflow of bile, loss of appetite, and too much fat. Similarly it also penetrates from the stomach to the liver, sets in motion and expels the gravel and similar diseases which exist daily in this air, purges, cleans out upwards and downwards, also the lungs, prevents pleuresis, the falling sickness, spasms, and the cramp.

Furthermore observe too concerning alum, that it also rises into the air and is received. Although its balsamic character is not as great, it still has the property of salt, and with it also that of healing whatever prepares to open, it also prevents the heated fluxes which produce devouring ulcers, so that they do not enter into the external channels. Know therefore concerning these three, what the nature of their airs is, and that they appear more commendable than harmful.

The second chapter.

Now know further concerning the three, that when they become air they are also not harmful externally, only through long use do they become too hot for the eyes, also for the ears. For such is their character that they cure many kinds of external diseases if their air is administered and used moderately and at the right time. Thus every scabies is cured by these three salts according to the order of the concordance, as the disease and the physic belong together. Thus the air of alum heals pruritus, the air of vitriol alopecia, the air of salt scabies. Moreover they also effect the healing of open injuries, internally and externally, like a wound-drink together with a good external salve, which is prescribed for an ulceration with its contrarium. Therefore there is not much to write concerning their poison; in the matter of temperature they are very close to the salt, which is suitable to all complexes, qualities and humors. Know therefore concerning the handicrafts that deal with the salt, that wet themselves with it, bathe with it, or use it externally in some other way, that it consumes all fluxes and other moistures of the body (unless some special cause exists there). Therefore the trades that use such materials, such as dyers, soapboilers and similar ones, are also the healthiest ones and do not easily become sick without some special cause.

On the other hand, now that the beneficial effect of these three has been reported, it is also proper to write concerning the poison which can be perceived here, and in the following manner. Every ore, that is not purified, but is still impure, contains poison within itself. Thus vitriol in its ore produces a mangy, scabby and itching condition, and draws the estiomenic fluxes out of the body. The salt with its impurity produces pointed scabs, also the trembling sickness and bad teeth, and alum, together with its earth, produces everything that alumen plumosum [9] does; but they do not cause anything harmful inwardly in the body. Thus too, the diseases that they produce externally are not full of great evil, nor are they essentially bad and evil as they show themselves externally. However, as soon as they arise they are as has been indicated above. Moreover know too that there are many species of salt, such as saltpeter and the like, also many kinds of alum. However, they should all be understood in the same fashion, and nothing should be considered in them except their strength which is contained in a graduated manner in them, and according to this should they be administered.

The third chapter.

In order that the things which concern the salt may be explained sufficiently, it is also proper to speak of the sorters. For the pieces which they use are taken from the three salts mentioned above, instead of from the common salt, saltpeter. And although the saltpeter is not saltpeter but saltniter, which is

---

[9] Talcum.

boiled and made from nitre, nevertheless it belongs to the mineral growths, and therefore it is correct to elucidate its species in the same manner as has been mentioned concerning the salts. Now it should be observed further, how aquae fortes, aquae gradationis, aquae regis and the like, as they may be called are made to a greater or lesser extent out of the three. What these can do to man may be described as follows: In the first place a compositum arises there, this compositum breaks the nature of its own united things so that there is nothing more to be said concerning each separate one, but only about the content of this composition. But since there are several kinds, such as aqua fortis, which is made from vitriol and saltpeter or from alum and saltpeter or with saltpeter and calcined alum, know then that the strength of the water must be considered, for the spiritus are its strength. They are that which changes the air. Thus the most subtile part of the spiritus comes into the air which we inspire. Concerning the same air, observe, therefore how it comes into us and how it acts in us. In the same manner as a vinum correctum which is poured together with the aquafort, it precipitates its spiritus like a red scarlet and separates the strength from the weakness. Therefore understand also that as soon as an aquafortic spiritus enters the internal regions of man, its spiritus are precipitated and weakened so that the power which they received during distillation is no longer found in them. Now since their power is weakened by the natural moisture, it is not necessary to know any more here, except the consumption of the precipitated spiritus. For they do not allow putrefaction, moreover they also expel themselves by means of their own expulsive property (virtus expulsiva). Now further, however, since these graduation parts are added as cinnobar, plumosum, verdigris etc., this is to be understood concerning them, that these same graduation pieces are not precipitated, but remain in their uselessness, the cinnobar like a sulphur, the plumosum like a fixed salt, the verdigris like a fixed vinegar. Now because they remain fixed, note then, that they do not come out of the body without harm. This, however, is their action, they penetrate so intensely through the brain, the lungs and the stomach, that they lose their natural forces, so that they no longer have their old digestions and expulsions. Experience teaches the recognition of whatever arises further from them according to the particular condition and property.

The fourth chapter.

As has been reported above many things are also understood concerning the aquis regum. But since they are made with other additions than with egg shell, than with the white of egg and the like, they become weaker than the previous ones are, but they can still be recognized alongside of the graduated water. Thus it is also to be understood here, that in distillation various things must be considered besides the salts, from which one must protect ones self more than from the salts, such as correctio mellis, correctio tartari and the like. These do not allow themselves to be removed from the spot where they have attached themselves. The diseases that they produce have special characters, signs, and properties, which can be recognized through experience.

Thus you should also understand the other distillations in the same way, because the ultimate matter shows itself there. For the virtues and properties which this exhibits are also true of all salts and distilled things. For it happens often that a thing is good but becomes a dangerous poison in the distillation, such as honey or salt; for the smoke of the salt produces changed, burned blood in those who smelt with it. Thus the salts, which they add to the ore and which melt so secretly also do this.

Such things also happen to the glaziers and those like them, the goldsmiths, the minters, who seek and make such additions for the purpose of flexibility, fluidity, gradation, cement and the like, not only in general but also in other particulars. Thus also to the alchemists, who seek many things in such materials and assay as a whole, in part and the like. In all of which the things must be experienced with which they asso-

ciate and according to which one must act. There are, however, many alchemists who reach old age; from this it may be
supposed, that the spiritus of the things probably bring them
there. I, however, cannot believe it but ascribe it more to their
abstinence, hunger, their mode of life, training and the diseases, which make an old person by themselves. If good
spiritus are added, as is reported, then their health is all the
greater. But those, who are not temperate in eating, also do
not arrive there; for these things require a good regulation.

The fifth chapter.

It is further necessary to speak of such workers, as are
occupied with sublimation; they cannot maintain themselves
without harm, and yet hunger and work are the best consumption of these things. Know at the same time also that the
sublimations have several good properties. Thus sublimated
mercury has the property that its air purges; now for one who
needs a purgative it is useful. Similarly the sublimation of
arsenic yields a heated spiritus in its air; by means of such
heat quartana is also cured, also some acute diseases, also some
fluxes, namely those of podagra, of arthetica and other fluxes,
which the wood guaiac can also cure. Thus in the air mercurii there are also all the virtues mercurii; for this reason,
those who carry on sublimation are not tormented by pustules.
Thus there are also precipitations, reverberations, calcinations,
roasting and burning and many forms of preparation, in all of
which an air spirit arises which is then of the same nature
and has the same properties as that from which it arises. Now
how these and the laboratory worker fit together, that is a
problem for the physicians, who live about them and experience
this daily.

Now understand therefore in this tractate, that the salts
exist and are used in various ways, but that they are not as
poisonous as the metals. Therefore a physician, who wants
to have a knowledge of these things, should gather experience
in this mineral school, concerning which nothing can be said

or written without a great and evident knowledge of the things. Even here nothing can be learned unless the pupils and those, who want to learn there, have worked in the mineral world and were raised there, without which it is not possible to discover these things. But as regards the knowledge of these things which is given to the blind (unexperienced), who can say that that which was painted is actually that which should have been painted; it always has defects, wants, and mistakes, which then create dislike. Man is made by man; thus nothing becomes perfect by reading or painting from something else, nothing tested, nothing confirmed, unless it arises from the foundation from which the river comes, where those drink who are thirsty for it.

THE FOURTH TRACTATE

ON THE CURE

The first chapter.

Now after the origin and the extraction of the mineral diseases have been explained in the first three tractates, so now it is not necessary to relate that the cure will be discussed in the fourth tractate. Thus there are two kinds of cure: one of the metallic minerals and the other of the salt. Know therefore that these differ from all others, and must have a special treatment. First, as far as the metallic fixed sulphur, sal, mercurius are concerned and those that are not considered as fixed, they need but a single description of their cure. What, however, must be known concerning the salt, is almost physic itself. There is more to point out however, concerning the trades and artists that are infected with such fumes. However in all that, which was explained, it is proper to attack the physic in another way, so that the fixed and subtile spiritus in their arcana will be conquered.

The second chapter.

In order to learn how to prepare the prescriptions for those diseases which have a mineral origin, know first, that of the air which was discussed in this book only the element fire shall be considered, since its elemental action is accomplished here at this spot. In the same manner as you may observe that a fire consumes like an element and not like a warmth, or that the sun dries not like a fire but like a warm thing. This warmth and fire are of two kinds, the one is an element, the other a quality.

Now there is nothing to be said here about the quality, only the element is to be discussed here. In the same manner as a fire works in the wood, notice then thus these spiritus act in the member to which they have attached themselves. And that it does not burn as rapidly as the wood is due to the living power which has a growing moisture within itself, by means of which the fire is deprived of its violence, so long until the moisture stops growing.

The third chapter.

Now it follows from this that the physic should oppose this fire through the power of the element, just as water alone extinguishes fire and nothing else. For the actual elements must be overcome by effective elements. For the quality and the element are different, that is, warmth is one thing and fire is something else. Thus it is not necessary to pay attention here to the warmth, but rather to the fire; for this reason complexion cannot come to complexion, since fire is not overcome by the complexions, but the complexion overcomes its equals and the element its equals. Thus for example: Suppose there were a disease which was warm in itself, now this warmth must have a moisture in itself or a dryness, which is innate as its diathesis, as De gradibus et complexionibus are understood. Now it follows from this that the physic too should be made by means of such a diathesis. Here, however, the things that concern the element are not to be considered. For in the element there is no diathesis, instead a united nature is master of the others, as the wetness of water is master of the fire and neither cold nor warmth. For this reason the water is wet and neither cold nor warm. The cold which is ascribed to it is an accidental coldness, which neither remains nor is helpful in the action of the wetness. In the same manner the cold is foreign to the water, as for instance when you cause it to boil over the fire. When the fire is no longer there, it returns to its temperature; thus too when the strange cold leaves it, it returns to the same middle temperature as after the boiling.

The fourth chapter.

Know then, that the disease is an element and not a quality and complexion. Now since the disease is an element, namely the fire, it is also necessary to differentiate the physic and to recognize it thus, that in it the element is wetness and not the complexion or qualitas humiditatis. For wetness and moisture are of two kinds and have these differences: whatever is wet is not consumed, but what is moist, that can become dry; what is dry can become moist, but it can do nothing against the element, and what is not consumable should be recognized as follows: whatever is subject to coagulation is moist, but that which is not subject to coagulation, is wet. In order for the things to be recognized I recommend them to one who is experienced in the transmutations and preparations of the art. Now because the fire and the wetness are in man, the wetness is not coagulated. However, as soon as the humida are administered against such a fire, they coagulate; now one dryness comes to the other and wood is added to the fire. But whatever is wet, even if it is distilled by the fire to a steam, yet it remains unbroken, for it redistills itself because man bears an alembic within himself.

The fifth chapter.

Now to speak further concerning the physic, know then that there is yet another power in it, the power of the elements which is overlooked in natural things. If it is a wetness, then the dropsy arises, which otherwise has no other origin; if the dryness is powerful, then ethica and its species follows. Then one can say: you have spoiled him by purging, that has happened because the element was disregarded, thus one can also say, you have overlooked the element fire and brought the patient to marasmus. Thus too if the element air is overlooked, you bring the sick one to colics and contractions. Thus too, if the element of the earth is overlooked, the patient falls into quartan exaltations (quartan fever). Now as the four diseases are elemental, not humoral, therefore it is also proper to write concerning their monarchy and to proceed according to it. Thus it is indicated, that the remedy should also be kept in this and should not be exceeded further, so that the aforementioned diseases should not follow. The sum is therefore that the mineral disease is a material fire which keeps its specific nature in its own world. For this reason the physic must also be a metallic wetness as is said above, and thus by means of the physic and through this process humidum radicale is preserved in its quantity. And, as long as it remains, no disease can be perceived; the prescriptions of this elemental wetness follow later.

The sixth chapter.

Therefore the remedies must be selected, which are free of coagulation. For example take alum, in this is wetness and coagulation. Now when they are separated from each other, the quality comes to one place and the same happens to the element. Now the element of alum is closest to the element water; for the element water also stands in its vessel like alum after its boiling, and after having been separated from the coagulated substances and having entered into its own clear element, but it is then deprived of the curative arcana of which the alum is not deprived. For the water is only a physic against the microcosmic fire. Therefore, it is only because of this that the wateriness is taken from the alum, and is so rectified that it becomes similar to sugar, and it should be drunk in the quantity of

1 scrupel. And if the signs of the element of diseases are perceived again, they are to be extinguished as before. Although there are more of such arcana, still for these I recommend the vulcanic school, since the things are to be learned there, which cannot be experienced here.

The seventh chapter.

Besides the arcana there are still some simplicia, which resemble the element water after the coagulation has been taken from them. Of these this chapter now treats, namely the water marrubii,[10] the water barbae Jovis,[11] also betonicae [12] and nenupharis.[13] Because, however, the art of such separation is not with the apothecaries, there follow instructions, how they are to be prepared, so that the workers who are set on fire by such spiritus may be maintained in health, the same too of those who are called alchemists, also the smelters and for those who have been mentioned in these tractates may it be salutary and helpful. And the instructions for cooking are thus, the herbs mentioned or their like are pounded while green and mixed with just as much cream of milk and boiled in a can for one hour in balneum maris, and then eaten in a sober condition. This food protects and conserves, defends and cures the alchemists, goldsmiths, minters, smelters and those, who were mentioned in this book. Pay attention to this that one half should be the component cream of the milk, or at least a fat milk, the other half the herbs as has been mentioned, or their like, as they are or the juice, although they can be cooked alone and then be cooked with the cream of milk as is indicated above; without this process the mineral diseases cannot be overcome.

[10] White hoarhound.
[11] Sempervivum.
[12] Pfaffenroehrlein (Betony).
[13] Water lily.

THE THIRD BOOK OF THE MINERS' DISEASES

in which alone the quicksilver diseases are comprehended.

THE FIRST TRACTATE

The first chapter.

In order that the miners' diseases can be comprehended in one work, there now follows the third book in which all the diseases that arise from quicksilver are gathered together according to their special properties as shown by experience. For the other previously mentioned diseases also required their own books for themselves. So know here at the beginning of this book that the disease of quicksilver has nothing in common with the sulphuric, mercurial and saline natures, nor is it connected with other metals or ores. It is also not to be considered here, that it is separated from mercurius, as if one wanted to say the evil which is separated from the good causes the diseases. For thus is the quicksilver within itself, it contains the good and the evil united, so that they are not to be separated from each other. From this it follows that whatever evil happens to you is to be ascribed to both the good and the evil, also whatever good happens is to be ascribed to both of them. To understand this take the example of theriac whose goodness without poison is of no use. Therefore mercurius is a born theriac, which lacks nothing, except the preparation to change it into theriac. Following this, know then, that the mercurial poisoning when it occurs through the pure metal of mercurii is a hard and serious impression (effect). For the things which arise from a hard and sharp impression are stronger in their malignity than the others. At the same time every thing which achieves perfection is without poison and well tempered, but it is not true of that which does not achieve it, which stops at a time when perfection cannot yet arise. These things are of a different nature than the perfect ones, and they should be watched more in good and evil things, than those whose development has reached its end. Thus the mercurius is also a half-growth which has not arrived at its perfection, but as a half-growth it achieves the same power, which it should have if it were to grow into a state of perfection.

The second chapter.

Now that so much harm arises from mercurius is due only to the fact that it has not reached perfection, so understand it thus. The perfection of mercurius is coagulatio, and because coagulatio is not there, it is a half-growth. Whatever resembles it, is to be understood like mercurius. Every coagulated metal has in it the type of mercurius, but since it is coagulated, therefore the same power is no longer present. For in the half-growth state and until the half-growth state is reached all metals are a mercurius. After this, however, follows the action to perfection, in which the separation of the metals takes place, each proceeding in the direction where it is directed. From this it follows that every metal can arise from the argentum vivum by means of the vulcanic fire, as it is found and seen in its origins. These things are pointed out so that you can recognize the mercurius, and can notice the difference between the coagulated and the uncoagulated. For all the things that are not coagulated and whose end and ultima materia should indeed be coagulatio, have all the species and kinds of poison and physic within them. Now there is none, whose end is coagulatio, which cannot be brought within this group, except mercurius; for many are the liquida (fluids), but their final nature is not coagulatio, but liquidum. Therefore the difference should be noticed here, in speaking of the nature of mercury, that it is not born perfectly but that it retains within it the property and the living and essential type, which the other metals contain within themselves in a dead state. For gold is only a mercurius. But because it is coagulated, it no longer has the same type. And although it is yet there, nevertheless it is dead. Thus too with silver, copper and iron which all have the mercurial type in themselves. In the same manner one can drown in water, but when it is frozen this cannot happen; as therefore the evil is taken from the water by congelation, thus coagulation also takes away the evil from the mercurius.

The third chapter.

What other cause can there be that this metal does not come to its perfection, except that the physician should seek his materia medica in this liquid, without considering the evil which arises from it, but only the great art which lies hidden in it, to discover this which expels both its own malady and other evils. Now because of the art the materia metallorum has been retained as a materia of nature, from which the six perfect metals are to be made and at the same time to find the essentia and the arcana, which lie hidden in the six metals; I wish to leave this and recommend it to the school to which it belongs. For here my intention is only to discover the evil which arises from it, and also its cure for the evil which has arisen from it. In order that its liquidum and its coagulatio be well understood, know then, that it is intended to be coagulated and is on the way to coagulation, but it remains a liquidum. From the same kind a vapor arises like moisture from water. Although the water itself is not seen to be a vapor, it is still an invisible water. And regard similarly water which boils until nothing more remains in the vessel; now that which has boiled away is not considered as water, but only as a vapor. This vapor settles in a distillatorium and produces water again, so that it is seen that the vapor is water. Now it is also thus with mercurius, which gives off a vapor through power of the upper firmament, like the water which is drained from it. Thus this vapor of mercurii enters into the air and with the latter into man. Now mercurius is in man like the vapor of water in the air, which is inspired. And as water again turns to water, thus mercurius again becomes mercurius, so that its distillation is seen and recognized, which is not only stimulated by the outer firmament but also by means of its own fire which it bears within itself, as every liquidum is consumed within itself

through the power of its own consuming fire. Thus the mercurius also drives itself to the distillation in the hills, clefts,
veins in which it lies. Thus where it is lies hidden and closed,
there it distills itself; in the hills its vapor cannot penetrate
strongly through the earth. Therefore those who live in the
earth of these regions must sit amidst such vapors like one
who sits in a bath-house; it is as if he were to sit entirely
in quicksilver. For the vapors make him wet who sits in the
bath, just as if he sat in the water.

The fourth chapter.

Now there is not only in mercurio such a uselessness, which
poisons man so seriously, but also in the stones; if they were
not coagulated, who would remain on the earth without an
evil? That is, without disease? For if their mercurius were
to remain imperfect like that of the metals, these distillations
would poison the earth to such an extent that nothing healthy
would be able to grow out of it. However, the congelatio
lapidum (congealing of the stones) removes all that. In the
same manner you see that the rebis (human feces) is the most
miserable and greatest stink in the warmth, so that no one can
remain near it, until the cold coagulates and freezes it, by
means of which coagulation all the disgusting things are
removed. Thus it is also with the congelation of the stones,
which does not permit this uselessness to rule. And if there were
as much of the mercurius of the metals as of the stones, and if
it were to lie on the surface of the earth, then the earth would
not bear anything useful as a result of its distilled vapor, and
it would poison the people on the earth like those in the earth.
For the regions in the neighborhood of mercurius and in contact with the mercurial watch live in greater worry because
of diseases than other regions. It is also unhealthier to
live there where mercury is kept than where it is not kept.
Understand also that the metals without fire do not show
any poison; for the fire drives them into their fixed mercurium.
And although their cold is unhealthy, it can still be compared

to and understood like a frozen water. Thus too concerning the rocks. But other growths like wood, which are not so hard in their coagulation, are closer to its matter than metals and stones. Therefore the vapor which emanates from them is to be avoided so much because its liquida materia has not been led to the hardest coagulation. Thus too the herbs have even less coagulation, for which reason their vapor surpasses all others. Therefore they are also a goodly prima materia, which is not so repulsive to man as the metals.

THE SECOND TRACTATE

The first chapter.

Now if the things are to be experienced according to the correct cause, as it is proper, then it can not happen without the astronomical physica. For the physician is subject to all the natural arts and wisdom which all gather together, and when they have all been brought together, he is first permitted to practise and prescribe against the disease. Now know then that the half-growth of the natural things has been dealt with in the first tractate, among which one is mercurius vivus, that can not be made solid but which remains fluid. At the same time it should be known that every natural thing which is open like argentum vivum, is like a house which is not closed and into which each one may enter; thus that too. Mercurius vivus is also open so that every physician can take out what is in it. But it is not so in gold, silver, lead etc. The door is closed by coagulation, and many obstacles must be overcome until with the aid of the art of opening up, of dissolving etc. its first matter is found again. Since the mercurius vivus is open it needs only to be directed to the vulcanic preparation The things now remain standing still and it must be shown further what mercurius is in its terrestrial constellation; that is, the earth has its heaven within itself and its astra are the mineralia. Thus mercurius is also an astrum; how and what, that is therefore to be recognized.

The second chapter.

You see that the year is divided into 364 days and some extra minutes. The year consists of two halves, summer and winter, and thus one year follows the other until the end of the world, and the summer follows the winter etc. until the aforementioned end: such is the creation of the world and thus is its order, and that is the world as man sees it. Now there is in the earth another world with other constellations, dwelling and the like, a special world which we leave here now. Now know that in this world it is not more than one year from the first day of creation to the destruction of the creation. From this it follows now, that after this year the things which are still in the earth at this time still grow (whatever grows on the earth does not belong in the earth); understand it thus. The seeds of the metals and the minerals have been sowed in the earth. They have their fall and their harvest in order to sprout sooner or later according to the arrangement of the godly order. In the same manner we know that now come the violets, then the thyme, then the roses, then cherries, pears, nuts, grapes etc. so long until the year has passed and it was a year. Although one comes later than the other, thus here too, now gold blooms in a region, now silver, here iron, there lead etc., that is past, that is present and the other future, that one's spring is past, that one's May, the other's hay month etc. for those who were there at the beginning of the world, they have reached the gold and silver of the spring with the violets, their successors have taken the gold and silver with the clover and the crowfoot etc. and so further and further from the first to the last the time of the year is divided. Thus what falls off grows no more, it is finished; no other year comes into the earth any more. And just enough grows on the earth, corn, fruit, grass etc. Thus man is also provided with metals, yet the difference is that this metal is that of the violet, this is that of the globe flower, that of the cherry, that of the pears, that of the corn, that of the grapes, that is according to the period of the year, and according to the months, which will follow each other for many thousands of years.

The third chapter.

Now since there is only one year in the world of the earth, as is indicated above, the division is as follows; know then that this year must also have its summer and its winter. Now summer on the earth has a severe warmth and winter is also severe and yet both are not palpable, but are a chaos. Here, however, in the world of the earth both are palpable, substantial, neither cold nor warm. They are cold and warm like the sun and like the moon, but closed.

Therefore observe the first explanation of the outer summer and winter, which is thus. There is a group of stars in heaven which make summer, among which the sun is the supreme one, which indeed makes no summer by itself but together with the band which belongs to it. Thus winter is also a gathering of stars which have their exaltations after the summer stars. And thus summer is a group of stars and winter also a group of stars, both being arranged to follow each other. Now observe then that the summer of the earth is a mass of ore pieces, which have the type of summer in their nature. And thus there is also an ore, which has the nature of winter within itself; and thus summer and winter are in the ore and not in the time nor in the day, or month or hour. Just as snow on a mountain and a ripe, blooming garden can exist next to each other, thus summer and winter also exist next to each other When they are opened and led into the creation, then the summer in the summer nature finds itself like the outer summer and the winter is like the outer winter in its nature; the property and quality can be compared, from which comparison the philosophy of the astronomical physician takes its origin.

The fourth chapter.

Thus there is only one year in the earth, and the summer and the winter are separated from each other, hidden and kept in their bodies, but they are not hidden in the external world because there these things are so constituted that their nature leaves them. But those that remain enclosed in them do not give off any cold or warm rays, and the exuding power of cold and warmth remains enclosed, although other properties burst and leave. In the same manner a secret kind radiates from the moon which vexes man, although it does not rule the cold with its stars. Thus this winter of the earth is only a mercurius vivus, that is an argentum vivum, and it has all the properties of the moon and the entire nature of winter. In it lie all the winter stars, the types of winter and whatever belongs to it. Therefore it is proper to speak of and to discuss the nature of mercurius vivus so that you may recognize what force and power it possesses, which cannot be understood without the aforementioned explanation; and although the name is not commonly used at this place and has not been chosen according to nature, but it has been named mercurius, like aurum the sun. Thus it has not been arranged wisely nor in accord with a well-founded philosophy. For mercurius, in accordance with its nature would better have been named luna, hiems nox, frigus or glacies. Yet because it is not firm, but is an open metal, therefore its name remains. Still, here it should be understood as luna or hiems, which corresponds to many stars. And silver should be saturnine and lead in place of mercurius, an order which the stars maintain and also the nature of things, as experience teaches their properties, nature and qualities, without which nothing can be treated nor discussed.

The fifth chapter.

Now that you have knowledge of mercurius, namely that it is a planet of the earth and luna, as the astronomy of the earth also proves, then know now that man is subject to the same planets. You know what power luna has over man and why the winter stars, which make the winter, bring man into their exaltations. Thus it is also to be understood concerning the nether planets of the earth, that they act in the same way on their climates and exhibit the influence of their power, but not over us men, for they are compact, coagulated, congealed, indurated. Therefore the actions which are in them cannot leave them, and so man remains safe from their action. But mercurius, which is called luna, is open, is not compact, nor indurated, therefore the lunatic effects can pass out of it as they do from luna, except the essential cold. Now since argentum vivum alone among all the stars, also on the earth, has an effect on man, know therefore that you should understand its nature and properties from the supreme luna, and what it is, that is mercurius vivus also; it is the luna of the earth, but with this difference: we bear the moon in the heavens only to a definite weight and in a definite degree, no more and no less, but only according to its course. The terrestrial luna possesses the same property, it can be absent, it can be present to a great or lesser degree or in excess. According to whether it is present, thus its effect may also be expected. Now this is the reason why we write about those who are hurt by the terrestrial moon, because they live, walk and deal with it. This cannot be compared with the upper moon; it is the cause of the lunatic diseases, concerning which I want to relate my experiences. For if it were possible for man to build a palpable moon, similar diseases would be found both in the upper moon and in mercurius.

THE THIRD TRACTATE

The first chapter.

To finish this project it is necessary to know how repulsive argentum vivum is for man, who can hurt or spoil neither the moon nor argentum vivum; they always remain uninjured in their nature. But man is the cause that man is made of the limbus, and this limbus has all the elements and natures within itself. Now since the limbus is such and man arises from it, it is therefore his father. From which it follows that the sun must fear the father and that the father commands the sun. Thus it follows from this, that we drown, that we must be poisoned by the stars, by the elements etc. and that we must succumb to their nature For just as little as a child can say, my blood is not from my father and mother, so we too cannot say that we can live without the elements. And just as a child is born of a father and a mother and retains the nature with which it was born until death, thus in us too our father's limbus must be recognized, which rules in us always, just as with a child in the mother's body. For all of us, although we are grown up, still lie in the mother and the uterus sur- rounds us all and whatever lies in the uterus must be the same as that which the uterus is and bears. Thus all the elements and everything that has been generated are around us. And we walk and wander among them. Therefore we are as loose and soft as a chick in its shell; for all the rays of the planets and of the limbus, which is the seed, enter into us and produce an essential action, for heaven and earth are the uterus and both are one thing. And man is the least thing and yet everything.

The second chapter.

At the same time, it should also be known, that two repulsive elements cannot meet each other without injury, just as the summer must yield to the winter and the winter to the summer. Now nature has created things so, that repulsive things do not meet. That is, repulsive things do not meet in the course of nature but only like with like. That is, green does not injure a cold herb nor a warm one, nor does the yellow, red etc. color, and the element remains without anything repulsive in its body. Yet it is said, however, that thunder, lightning, showers arise from repulsive things, but it is not true. For the noise, the light and the fire come through the power of the elements and from its own nature and not from repulsive things.

For see that gunpowder is sulphur and saltpeter, they are both one element and yet a repulsiveness is present This repulsiveness is not in the element but in the fluid. For the fluid of sulphur and the fluid of saltpeter can not be together alone because of the fluid, and not because of the elements. Thus it is also with the origin of thunder. Just as a man and a woman both have a warm complexion, both are of one nature, property and quality, as far as blood, flesh and bones are concerned. But that they do not remain together, but flee from each other, is not the fault of nature and of the element, but of a nature, which is like a fluid, which differs in form and application. This application is the contrarium, which shall be discussed. Thus through such an application repulsive planets become violent and the like; this I leave to other books.

The third chapter.

Now observe as follows that two repulsive things cannot remain in their element without becoming disorganized. Man has a warm nature, and every human being is like unto another in warmth, both men and women. That one says, however, he is cold, the other is such etc., is false and without foundation. In the same way a woman also remains in the warmth of a man and does not become colder. From this it follows, since man is warm by nature and the terrestrial moon is cold, the two are repulsive to each other. But why should the cold of mercurii concern man? There is no reason for this, for they do not come together in one object nor are they forced together; therefore they cannot break one another. Here, however, we are concerned with the harm which is to be explained and which is a tincture; the element itself is not present but it acts upon man through its tincture. Thus just as when one throws a poison into water, the two are not repulsive, for the water does not oppose the poison, for there is nothing in it which opposes the one or the other, but rather its warmth etc. Know therefore that argentum vivum has a poison within itself. With the poison it tints man, and the tincture, which it and other planets have in an invisible form, and which the dyer has in a visible form, is the tincture which produces sickness. Therefore it is not that one element breaks another in its element, but both remain in their elements, man warm, the tincture cold. And since the external tincture places the cold in man, therefore the disease is there, and health and the diseases can well remain together, nor do they separate, but both remain wholly and completely together. Therefore when the disease is removed, health alone is present. If it had not remained during the disease, then the patient would not have recovered. Therefore they are well able to remain together, but whatever rules, that is effective.

The fourth chapter.

Now since mercurius vivus is a winter, thus it also acts like the winter. Winter does not drive the warmth out of man's body, but it dulls it, something which it cannot stand. And when one freezes, the cause is not that the warmth has been expelled from him, for it still lies in his center whither it has fled in a compact form. He dies, however, because the warmth has ignited the heart too intensely, and the warmth leaves it and permits the cold to enter, so that the heart suffocates. For warmth and cold do not mix into one in a born element, as wine and water are mixed, but like oil and water they remain separated, and thus the cold also remains. Thus it is also with the luna of the earth; its cold is not evident but the signs are noticeable, therefore mercurius vivus produces shivering, chattering of the teeth and the like; if it were not the terrestrial winter, it would not be able to do this. For every form of shivering is an influence of the effect of the winter stars. And although warmth and warmth also produce shivering, also cold and cold; still that does not belong to the mine and earth diseases, therefore the cause will remain reserved for another place. Here it is my intention to describe only the mineral diseases of the earth, which arise from their stars, still there is no need to go any further than to describe mercurius, as it has happened; for it is the terrestrial luna, the terrestrial winter. And thus it has two effects on man, one as a moon, because man is made subject to it, the other as a winter, because of the same stars and a like nature. How all that is to be recognized need not be described here, because the origin of the diseases has been indicated and sufficiently explained. It is therefore my further intention to heal the same.

THE FOURTH TRACTATE

The first chapter.

Since it is now proper to instruct regarding the cure, know then that the winter which is in mercurius produces one kind of disease, which is a shivering without any frost being felt, as it has been reported. This you should understand as follows. Each frost is a mortal disease which like freezing drives all the heat inwards and the latter consumes the heart, and wherever the heat disappears there the member freezes. Now it cannot retreat from the center and cannot flee further and therefore it consumes it. Such also happens to the shivering one due to mercurius, the lungs, the liver, the stomach, the brain etc. all burn internally. Where they shake, there the heat has left them. Because of this retreat the heat in the center becomes too great. Now a fire arises there; this consumes and acts according to the species and nature of its element. Now there arises putrefaction of the liver, putrefaction of the stomach, putrefaction of the brain, putrefaction of the kidneys, putrefaction of the viscera and the like and many such diseases so that they cannot all be mentioned. Then too the marrow is consumed in the bones, the blood vessels, the bones, the blood, the flesh in the skin, the cartilages and whatever else is in man. At the same time other diseases also arise, which are peculiar to the species and properties of such members, such as the diseases which belong to the lungs: heavy breathing, short-windedness, coughing, ulcer, putrefaction, consumption etc., to the brain: mania, frenzy, headache, fluxes, toothache, paralysis, apoplexy, lethargy and the like, whatever is comprehended there, and thus it is with all the other members, which it is not necessary to relate here since it is sufficiently apparent to all physicians what diseases may be produced by and arise out of this cold.

The second chapter.

Now, however, enough has been said concerning the diseases which mercurius produces by its cold; those who make themselves subject to it accept it externally like the odor of a rose. Now further concerning the diseases which it produces as a luna (moon); know concerning it that the luna is also a new, quarter and full moon and that it waxes and wanes. Now it is thus. It grows from the birth of Adam to half of a man's face, i. e. of the world, and if the half of the world is present, then it is its full moon, the waning is then the other half. And whoever does not recognize this moon, he speaks in error concerning the age of the world or of the judgment day. However, what its full moon or quarter moon are like know from this. You know the nature of the new moon as it concerns man for the upper heaven teaches that; you know too how it waxes to a quarter, and then from a quarter to the full moon, then it passes from the full moon to the last quarter and from the last quarter until it vanishes. Now as you know from astronomical experience, even if you did not have the ephemerides, but only the lunatic diseases, still you would know how the moon stood by the signs that you would recognize in the sick person; for the lunatic diseases are certainly ephemerides of the moon. Now, if without knowledge of the moon's course, the moon's course is found in the sick person as in a calendar, then note that the mercurial diseases are found in miners with the signs of rising or with the signs of setting; recognize thus whether mercurius is rising or setting. If God would decree the end of my archidoxes then you would have to learn about nature in a fundamentally different way and thoroughly; therefore open your eyes, if you want to proceed correctly.

The third chapter.

Now if the moon mercurius causes diseases through its moon action, then know the signs of some of them in this manner. It is said that the brain is the member of the moon in man,

i. e. the brain is the luna of the microcosm and rules in him. Know that you should not seek this cause, for the moon acts on all the members; not on one alone but on all. Thus the brain is also like any other member of man. The fact, however, that the brain is attacked more in the average man, that alone produces lunacy. Thus if the brain suffers, it is noticed and felt in the reason, this waxes and wanes like the moon. Since this is so easily understood and is comprehensible to the average man, therefore it is said that they are similar. Just as it is known concerning reason, thus it should also be known that such signs are to be found in all the members and not only in the brain. Now the lunatic diseases are frenzy, madness, catbite, mania, St. Vitus dance, the falling sickness and the like, both chronic and acute. Now the reported diseases, which are mortal or appear together with mortal diseases and attack the brain and the region of the head, are paralysis, gutta, arthetica, podagra and its species. If the liver is attacked, intermittent fevers, jaundice and similar diseases appear, and if the kidneys are attacked, diabetes and its species appears, and thus also with the other members. Now they are mentioned here so that you should know which diseases the moon rules, for mercurius also controls the same ones corresponding to its exaltations, whether it be dropsy, consumption, ethica, quartana etc. These things are understood with the aid of the external astronomy, which it is not necessary to describe here, since it has been sufficiently explained that the terrestrial moon would be considered here and not that of heaven, both in the cure and in the recognition of the disease.

The fourth chapter.

Now it has been said that mercurius can produce external diseases through the action of its vapor and through hidden rays; know too that it also acts in the same manner when it is taken as physic. It would therefore be good if the mercurial physicians who prescribe mercurial medicines in the form of salves, fumes, precipitate, corrosive water and the like, would be better instructed concerning the nature of mercurius and would reflect upon it ⟨so that they would not succumb to any imposture⟩. For mercurius is an eternal luna, in which there is no death until the judgment day, be it that it enters into man or remains outside, at the same time it is also a permanent, constant and strong winter whose snow no sun can melt and whose ice cannot be thawed. Now since such can be recognized here, it would be well to attack it with greater understanding. It is also apparent that man cannot digest it, that he cannot consume it; for all metals are digested, but not this one; for neither man nor the ostrich is capable of digesting it. But, however that may be, mercurius remains alive and does not die, and if you place it in the body, the latter has this luna within itself and also its property. Now to this is added the fact that the external moon unites with its own kin and carries on its government besides the first one more joyfully and thus produces here these double diseases, ⟨for what the moon is capable of doing in ruling over all diseases⟩, that is also possible for mercurius. Serious diseases result when it is used in a wrong manner. If this luna mercurii is not mastered, then all the diseases in it are incurable; unless this moon is subdued, nothing helps. For the correct cause of the physica, to recognize and to do the things. . . .

. . . may be recognized as an example of the miner's diseases and of the things, which develop in them, i. e. in the mines, where such diseases are.

The sixth chapter.

Thus enough has been said concerning the preservativis and the conservativis; for whatever conserves also preserves, and whatever preserves also conserves. Now the cure must be discussed according to the division contained in the last chapter of the 6th tractate in the 3rd book. Mercurius vivus must be removed from the body before the curative physic can begin to act. If this is to happen mercurius must be alive, and as long as it does not become or is not alive, it cannot be removed from the body. Therefore it is most important to make it alive, and then to expel it from the body. Note that the process proceeds in this manner. Assume that you had a sick man who bore a certain quantity of mercurius within his body and indeed live mercurius, then he would exhibit the following signs: the teeth would be almost black, the limbs lame, and the disease would wander from one place to another, first of all in the limbs and joints. These are the signs of living mercurius and usually at the same time they are at a fixed place, as if a hard abscess lay hidden there. However, if mercurius is not alive then the disease corresponds to the moon, with hardening of the bones, the urine is discolored, and the breath stinks. Thus both mercurii are recognized and understood.

The seventh chapter.

Now it is proper to speak further concerning the evacuation of mercurius which is not alive, and how to make it alive and apt for evacuation. Note the following concerning evacuation. Every mercurius settles in the cavities of the joints. That which passes through the backbone and the hip region falls into the knee joints or ankle points through the corresponding ligaments. In the same manner mercurius vivus passes downwards when it is placed in a trench until it finds a cataract, where it remains. Thus the knees are cataracts, also the ankle joints, the hip joints and the joints of the backbone. And sometimes it collects in the ligaments, sometimes at the bottom in the soles, as deeply as it can fall. Thus it also settles in the arms, in the cataracts of the shoulders, of the elbows and further into the wrists, sometimes it falls into the neck, sometimes out of the hollow of the corner of the eye, sometimes through the nares, frequently through the pharynx into the stomach and passes out with the stool. All this can be recognized where the things are present by means of the signs and good experience.

The eighth chapter.

Now if you find a cataract where you observe that mercurius has gathered even though until then it was believed that it was not there, then do something and place this corrosive over the entire area of the cataract, as extensive as the area may be, and do not pay any attention to any joints that are there, do not worry or be timorous on that account but make the corrosive thick and strong enough. The description of this corrosive is as follows:

Rec. realgaris albi ij half ounces
      alkali of lime and iron dross j half ounce
      oil of roses as much as is sufficient to spread it,
      as is indicated above.

It is the nature of this corrosive that it heats powerfully, and through his heat in the limb the mercurius begins to run and flees to the cataract. Following this, know then that you should carry out this treatment as long as possible, for the longer the better, namely for fourteen days or three weeks until the scab falls off of its own accord and the mercurius runs out. Then heal it with a rubber plaster until it is filled with flesh; then treat the area with crocus martis until it is covered with skin. At the same time you should also know that you must avoid other corrosives, namely sublimato mercurio and the substances which corrode quickly.

The ninth chapter.

The dead mercurius must be revived so that it can leave through this exit, notice therefore how it is revived. First prepare a strong water bath of herbs that are rich in mucus, of the buds of firs or juniper bushes and also of fresh fir cones that have been boiled; let the patient bathe in it as hot as he can after having considered his strength. The same can be done in the hot springs of Pfeffers, Baden, Plumbers, Gastein, Doeplitz, Ach etc. or Embs, Goeppingen etc. or in artificial sulphur baths etc. And when they come out of the bath they should be massaged with succo flammulae[1] or with oleo de piperibus,[2] following this they should be made to sweat, in any way that it can be done, either with theriac or with mithridate and the treatment should be continued without taking any notice of the patient's pains or days of sickness; for it becomes alive through such pains. When it then shows itself to be alive, which can be recognized with the aid of experience, then treat as is indicated above. And as it often occurs that the pains do not cease, then look to the center so that you can recognize it. For often the mercurius which has been prescribed in the physic, has been deadened to such a degree, because the ingredients do not fit together, that it only revives very slowly, which the pox physicians consider a great art, but which I think is a great folly.

The tenth chapter.

It is necessary to write further about the diseases, especially after the aforementioned treatment has been carried out; so note first of all how the shaking limbs, hands, feet or bodies are to be treated. And this process is as follows, first prepare a bath of agimonia, of floribus lilii convallium, of radice hirundinariae[3] and of a few parts of egg shell. Let him bathe in this bath and after the bath anoint him with this ointment: Rec. boiled fox fat 1 lb., add to it 10 half ounces of distilled castoreum and 15 half ounces of distilled turpentine with pepper, cantharides and baccis lauri; Mix this mixture over a slow fire and anoint the patient with it; you will hardly learn anything better, not only against mercurial shivering, but also against gutta, paralysis, lethargia, and also whatever is possible to be preserved in apoplexia whose origin is recognized as mercurial, or in mercurial cramps, is healed by this salve. At the same time it should also be mentioned that you can improve the bath with flammulae and sabina according to your

[1] Crowfoot juice.
[2] Pepper oil.
[3] Lily of the valley, and swallow-wort (celandine).

opinion, or that you can omit the bath altogether according to your judgment.

The eleventh chapter.

Note then, that the miners' jaundice should also be treated as follows, after mercurius has been removed, namely with two kinds of physic, mineral and mundane in the form of one part asula [4] and one part rebisola; [5] administer a drachm of these two every morning until the sickness vanishes. But better than this is extract of rhubarb with liquor tartari, used as experience teaches. Miners' dropsy must also be treated in the same way, also ascites and tympanites, also hernia, bubo, etc. together with the medicaments which are applied externally and which are very helpful in such natural diseases. At the same time, know also that this is a laxative for women, when they succumb to such mine diseases because of the uterus; it also helps in the fevers and in their colics, contractures and other cramps; for whatever the liquor tartari does not discover, it will not be possible to find it. It should also be noted that by bathing the contracture of mercurius with serpentina and with lilio convallium, and by continually anointing it with axungia humana medullata or vulpina medullata or the same from a badger and by diligently keeping it warm, every mercurial contracture will be opened, only do not despair because of the long time.

The twelfth chapter.

Furthermore, several accidental diseases which rage in the chief members, also occur without any other accompanying diseases. As for instance an especially intense stomach-ache or stitches in the side, in the spleen or the liver, or strong headaches, or kidney pains combined with intense pain in the back: these are not to be placed in the order of the prescriptions, but are to be treated with the great arcana, such as laudanum or the materiae perlatae and the like. For they are so hard and difficult to attack, that something can only be accomplished

[4] Wolf's milk.
[5] Probably diminutive form of ribes = currant bush.

by means of the powerful quinta essentia, for there one must act in the same manner as when water is poured over a fire and extinguishes it by force. Whatever needs such great power, must also be fought with great arcana, as element against element, complex against complex, gustum against gustum and the like.

The thirteenth chapter.

Toothaches are also present sometimes together with many accompanying symptoms, with blackening, putrefaction, loosening and loss of the teeth, and with strong stabbing pains; note the treatment as follows. There is nothing special to be said about the blackening since strong waters or tooth powder removes it; therefore I will not add anymore to it. The putrefaction, however, must be treated as follows: the teeth must be rinsed with honey water and then smeared with a mixture of honey and aloepaticum for several days, then they should be rinsed twice every day with distilled and prepared alum water, containing waybread salts, consolida, serpentina, etc. until the putrefaction of the teeth disappears. What can be better for loose teeth than to smear them with oleum de croco martis? For falling teeth there is nothing which removes the mercurial power better than oleum de nuce muscata. The stabbing pains, however, cannot be treated with poor mixtures, but with the arcana, as has been indicated concerning the chief members, and at the same time use bleeding and cupping as ordinary usage demands it.

The fourteenth chapter.

Now concerning the wounds caused by lightning, note that they alone are like a burn, and are not to be treated otherwise than with axungia porci (lard), which is melted very hot and poured into succum barbae and then made into a salve. For the same purpose, boil milk with crayfish and then apply it as a salve. And wherever wounds have to be healed or skin closed this should be done by means of the egg salve and the drying powder. Know further concerning the wind and the dragons which also occur in the mountains, that auricula muris should be added to all medicines because it has a special property to overcome and expel these things.

FRAGMENTS AND SKETCHES FOR THE 3 BOOKS
ON THE MINERS' DISEASES

The second chapter.

Now just as magical fires and winds are seen outside, thus they also occur in the mines. Now their origin is as follows: if a spirit is like man in its acts and deeds, it can also act like the latter; that is, a spirit can go into the fields, chop wood and carry on all its business like a human being. With this difference, however, that it does not split natural wood, nor does it carry on natural handicrafts, that is not with natural things, yet it can do it with supernatural things, where wood and handicrafts are also present, but these handicrafts have their special cloth, wood, etc. They are spirits, there is nothing substantial in them, therefore their fields and cloth are also of this nature. However, since man is natural, he must also have his natural tools. Now man can make a fire from natural things, therefore the spirits can also make a spirit fire from supernatural things, which is like them. For man makes a corporeal fire and his things are corporeal. From this it follows that they can produce roses, horses, human beings, flowers, etc., but such as they themselves are, immaterial and not corporeal. For it is know as the spirit world with its incorporeal things and the world is called corporeal with its corporeal things. Thus dragons and other shapes shoot up which are not natural, and thus too winds and the like, which are also not natural. Now where such phantasy of the spirits breaks out, there many such things occur, which have been sufficiently attested in their book. At the same time know too that it often happens that witches and sorceresses cause such things, since the spirits obey them. It does not occur because of their actions, but they only believe that they do it, still these are all spiritual things about which I do not want to dispute here, but only to indicate sufficiently that these things should be differentiated from the natural ones in the practise of healing. Therefore know that although they do not employ anything natural, still they injure that which is natural, i.e. they deceive that which is natural and vex it both externally and internally.

The third chapter.

Besides that, however, know that thunder, lightning and sheet-lightning occur naturally in the hills; therefore a short meteoric explanation concerning it. Sheet-lightning from the clouds is a vapor of sulphur, which ignites itself just as a fire is lit by a flint, as is explained in the meteoric writings. Thus there is also a sulphur vapor in the mines, which arises from the terrestrial stars and which can be ignited by the air, for otherwise it is only a glimmering fire. Following this, note now that if a sulphur vapor lies in a mine, it is just as if sulphur fumes would be created in a sublimatorium, but since no fire can reach the fumes, they do not burn. Now if a light, a fire, or a crucible is carried in by the miners and the vapor reaches it, then it burns just as if a fire would come together with air in a sublimatorium. And because it is only vapor and nothing else, it is soon burnt out and is comparable in its nature to sheet-lightning. For all the things that burn thus are sulphur, whether it be on the earth in this firmament or in the earth in this firmament. Thus it can also be explained that such sulphur produces a sulphur stone, which when ignited burns into the stone, as long as such solid sulphur is there, whether it be earthy, stony, margazitic, like talc or like bismuth. For there where the ores are not boiled enough and have no stable composition and fire ignites them, there the same thing happens to them as to a heap of coal, which is ignited by a coal and is completely burned. Therefore understand here too that such vapors are also present in the clefts, passages, galleries and caves of the mines, which burn when a light reaches them until the entire vapor is consumed and extinguished as is seen then.

∴

However, in order that you should know that mercurius is in the earth half the time, and half in the firmament, note then, that the earth is a special world, which has its special star within itself, and the ore is its star, and mercurius is an ore, therefore it is now a star of the time. That is, it is the winter, which the earth has. As you see the sun is the summer and is a star together with all the other summer stars, that is the part which is the summer; know then that the stars that are the winter are also similar, videlicet galaxa; as long as they remain so long does winter stay and so long is it cold etc. Thus know then that summer is also in the earth, and they are stars like the sun and its stars, thus the winter too, like galaxa and its stars. These stars are divided into two kinds; mercurius is one kind, that is the winter, and the other kind is the sulphur in the mines.

From this know now that the winter is half of the minera, the other half is the summer, half is the mercurius of which I speak here. For thus it is arranged that summer and winter lie near each other, but separated just as they follow each other on the earth; thus they are near each other there. All the winter stars (sidera hiemis) live in mercurius and all the summer stars in sulphur and remain thus near each other for ever and ever; this is the nature of the terrestrial star. Know further that winter lies in mercurio, and therefore causes shivering and freezing because of its nature, for there is the winter.

III

THE DISEASES THAT DEPRIVE MAN OF HIS REASON, SUCH AS ST. VITUS' DANCE, FALLING SICKNESS, MELANCHOLY, AND INSANITY, AND THEIR CORRECT TREATMENT

BY

THEOPHRASTUS VON HOHENHEIM
CALLED PARACELSUS

TRANSLATED FROM THE GERMAN, WITH AN INTRODUCTION

BY

GREGORY ZILBOORG

INTRODUCTION

The Place of Paracelsus in the History of Psychiatry

This book of Paracelsus is perhaps more recondite and diffuse than any of his medical writings. It can more easily be understood if one bears in mind the historical and cultural atmosphere in which Paracelsus lived. It is seemly to recall in this connection an episode from the life of Rabelais, the cultural as well as chronological contemporary of Paracelsus. Rabelais was at one time a member of the ambassadorial retinue which visited Pope Paul III. Rabelais " being bid to make demand [for some favor] only begged his holiness would be pleased to excommunicate him.

" So strange a request having caused much surprise, he was ordered to say why he made it. Then addressing himself to that pope, who was doubtless a great man, and had nothing of the moroseness of many others: ' May it please your holiness,' said he, ' I am a Frenchman, of a little town called Chinon, whose inhabitants are thought somewhat too subject to be thrown into a sort of unpleasant bonfires; and, indeed, a good number of honest men, and, amongst the rest, some of my relations, have been fairly burned there already. Now, would your holiness but excommunicate me, I should be sure never to burn. My reason is, that, passing through the Tarantese, where the cold was very great, in the way to this city, with my Lord Cardinal du Bellay, having reached a little hut, where an old woman lived, we prayed her to make a fire to warm us; but she burned all the straw of her bed to kindle a faggot, yet could not make it burn; so that at last after many imprecations, she cried, " Without doubt, this faggot was excommunicated by the pope's own mouth, since it will not burn." In short, we were obliged to go on without warming ourselves. Now, if it pleased your holiness but to excommunicate me thus, I might go safely to my country.' " [1]

[1] " The Life of Rabelais " in Urquhart and Motteux's translation of *The Works of Rabelais*. London, 1849, vol. I, p. 7.

Thus the psychological motto of Rabelais was, " I do not wish to belong." This was the motto of Paracelsus also. Both men belonged to a generation of which many individuals, cheerful or sad, angry or compliant, conservative or radical, would not comply with the dogma and tradition of the day, regardless of whether the dogma was religious or scientific. Luther vituperated against the Pope hardly more violently than Paracelsus against Galen. To challenge, to abandon, and actively to combat Galenism was a difficult task at the time. It was difficult not because of the power of intrenched conservatism and the vested interest of the medical profession. It was psychologically difficult, because there was no theoretical or empirical body of knowledge with which to counter the weight of tradition. This is particularly true of mental diseases, which were at the time a department of theology and not of medicine. They belonged to demonology and not to psychiatry, and the best textbook of mental disease, the most popular and the most authoritative, was the *Malleus Maleficarum,* and not a treatise on clinical medicine. The *Malleus* was willing to recognize the authority of Galen only for the purpose of emphasizing its own preconceptions; the result was that very few *natural* diseases of the mind were recognized. Whatever seeds of sound psychopathology there were in Galen were thoroughly sterilized by the prevailing demonological doctrines.

Paracelsus' opposition to Galenism was threefold. First of all it was intellectual. The Galenic psychophysiology was definite and systematized. Paracelsus, striving to create a biological concept of man and to synthesize this concept with that of everything living in nature, was unable to feel at home with Galen to whom man was but another anatomico-physiological apparatus. Then, too, Paracelsus belonged to a humanistic age; this aspect of his opposition to Galen was of a psychological nature. Paracelsus never lost sight of the patient as a human being; Galen was interested in the physiological aspects of the person, the so-called humors which as the ages went by became postulates rather than physiological phenomena. The *succus melancholicus* and the *black choler* were assumptions not to be questioned. Paracelsus as a true son of the sixteenth century would question anything which he was unable to relate directly to the man with whom he was dealing. There were, thirdly, sociological grounds on which Paracelsus felt the need of objecting to Galen. Paracelsus was a man of the people and of the earth, while Galenism was a kind of stamp of professional respectability, of being professionally well born. If we recall that the decisive opposition to Galen which Paracelsus displayed while professor at Basel represented almost a revolutionary attitude, an attitude that well fitted his temperament, it will become clear that when Paracelsus approached the problem of mental diseases he threw overboard a tradition which was not entirely abandoned until the eighteenth century. It was only abandoned when, almost two hundred years after Paracelsus, Georg Stahl and later those at the University of Montpellier (Boissier de Sauvages, Barthez) attempted to introduce the concepts of vitalism into the biological consideration of disease in general and mental disease in particular. The originality of Paracelsus thus stands out in even greater relief.

There is another point to be remembered in this connection. The book on mental diseases, the English text of which is here submitted, was written when Paracelsus was about thirty years old, that is to say, in the early twenties of the sixteenth century. Weyer's *De Praestigiis Daemonum* did not appear until 1563 in Latin and 1566 in the vernacular. In other words, Paracelsus anticipated the descriptive method in psychiatry by almost fifty years. He describes the clinical manifestations of epilepsy, mania, and hysteria, which he chooses to call *chorea lasciva*, in the manner of an earnest observer—a manner which does not appear in psychological medicine as a real methodological principle until the end of the eighteenth century and which was not established as such until early in the nineteenth by Pinel and, more particularly, his pupil Esquirol. It is indeed surprising and rather puzzling to find so many details, so accurately related and so well arranged, in these clinical descriptions of Paracelsus. It is surprising because this was not at all the customary approach of the physician of the time. Even Agrippa, who was Paracelsus' contemporary and who was particularly interested in the " witches " and the " bewitched," did not describe the various mental conditions with such clarity or detail. It is puzzling because there is a mystery as to when and how the thirty-year-old Paracelsus acquired his rich clinical experience in mental disease, for there is no doubt that without such experience it would have been impossible to render these descriptions.

It is also impossible to overlook the fact that, apparently unfettered by any of the established principles of the medical law and order of his day, Paracelsus was not afraid to admit that he was observing a new aspect of a disease whenever he stumbled upon one. His whole approach to chorea is illustrative of his temerity and intellectual freedom. He takes the credit for this disease away from St. Vitus with the rebellious enthusiasm of Hippocrates depriving the gods of their role in the formation of epilepsy. Moreover, historically he was the first to differentiate the sexual components and the unconscious factors in the development of hysteria. This he does, not by way of inference or philosophic speculation, but by direct observational assertion; he consequently suggests that the name of *chorea lasciva* be adopted and that the unconscious (*unwüssende*) suggestibility of children be considered. He refuses to follow the traditional assumption that children do not possess an imagination, and once he sees the manifestations of the latter in youngsters suffering from chorea he recognizes and asserts them with his characteristic lack of equivocation.

The theories of Paracelsus concerning mental diseases and their modes of treatment, interesting as they are, can be no more than mentioned here. A special study of these will be well repaid, but this is to be left to some future time when the English text will have been before the reader for a longer period. To understand these theories it is necessary to weed out the Paracelsian pharmaceutical divagations, which were more a reflection of the time than a product of strict originality. Too, the limitations of the language will have to be taken into consideration. Paracelsus lectured and wrote in the vernacular, leaving Latin to those of his pompous colleagues who were clad in silk and velvet and who were not brought up on cheese and similar coarse foods. The vernacular of the time possessed neither the terminology nor the flexibility requisite to the swift and volatile thought of Paracelsus; his conceptual thinking, both in depth and shading, was considerably ahead of his language. He speaks for instance of " natural " disease quite obviously meaning organic and physiological; when he speaks of " spiritual " diseases he quite obviously means psychological ones. There is no doubt that when Paracelsus emphasizes time and again that a disease may be caused by a disturbance originating in the *spiritus vitae* and that the *spiritus vitae* may produce physical symptoms, he is presenting, in a redundant and perhaps confusing way, the definite conception of psychological illness and of conversion symptoms. Even his discussion of *suffocatio intellectus,* in which he attempts to explain that reason might appear profoundly affected but that actually this disturbance might be but a result secondary to emotional or other psychological disturbance, is as keen a clinical observation as it is modern.

Apparently Paracelsus was not unaware of the fact that mental disease is a highly individual phenomenon and that in order to understand a given mental disease a high degree of individualization of one's point of view is needed. Therefore he constantly insists on the determination of the patient's " complexion " and on the variability of clinical manifestations of the same disease. In this effort to formulate a definite psychological attitude on the part of the physician more than in any other, the difficulties inherent in the cultural age of Paracelsus are evident. What was apparently in the mind of Paracelsus, what he intuitively perceived and conceived, was the concept of personality which was not formulated until the middle of the nineteenth century, particularly by the German psychiatrists of the romantic school. In the days of Paracelsus the personality was culturally born, but it was not yet subject to scientific analysis—it was lived instead. Paracelsus was one of those who *lived* his personality. He suffered from the same difficulty as his great contemporaries like Rabelais, Agrippa, Vives: they all became men of action or melancholy contemplation; they wrote a great deal with more temper than judgment. They were ahead of their time—a time which was not yet ripe to recognize the value of the individual and therefore still less to formulate the concept of personality. Paracelsus must have felt considerably ill at ease and bitter, for what he felt he knew was true; but he could not say it fully, and with a unique bitterness and rancor he resented the recalcitrant and unfeeling attitude of the medical profession toward the problems which engaged so much of his ardor. Consequently he felt shocked by contemporary medicine and with characteristic venom he repeatedly called it a whore for its venality, subservience to the powers that be, and complacent, uncritical hugging of traditions which had outlived their cultural and clinical usefulness.

The text submitted here is an English rendition of the Basel edition of 1567. This was the first edition of the book. Huser's first text appeared in 1576. Sudhoff preferred to use as a basis the later Huser text combined with the two mentioned above. From the standpoint of the general ideological principles and the fundamental clinical attitude there is no difference among any of these texts. The edition of 1567 was therefore chosen because it was the first to appear. It was deemed of some interest to render the prefatory dedication of Adam of Bodenstein who was responsible for the first publication of the book. Bodenstein's postscript, however, dated February 1567, was omitted here because it adds nothing new to the atmosphere in connection with the issuance of the book nor does it elucidate any of the views of Paracelsus evolved therein.

THE DISEASES THAT DEPRIVE MAN OF HIS REASON, SUCH AS ST. VITUS' DANCE, FALLING SICKNESS, MELANCHOLY, AND INSANITY, AND THEIR CORRECT TREATMENT

BY

THE MOST HIGHLY EXPERIENCED AND ILLUSTRIOUS PHYSICIAN

THEOPHRASTUS PARACELSUS

With a description of interesting and useful procedures, applications, and effects of vitriol and metals, compiled from books of the above-mentioned author, and faithfully published by Adam of Bodenstein.

Contents and findings of this book are duly outlined in the preface.

In the year 1567.

To the Reverend, Noble and Honorable Sir,

PHILIPS GEORG SCHENCKE ZU SCHWEINSBURGK,

> *Dean of the chapter of Fulda and Abbot of the*
> *monasteries of Holtzkirch, Neuberg, Dulba, and*
> *Saint Michael, etc., my magnanimous sponsor and*
> *dearly beloved brother-in-law.*

If a person wants to have a book published, it is an accepted, long-established custom for him to find a noble patron for the safety and protection of himself as well as of his writings. How much more necessary it is for very important publications to choose honest patrons who, in contemplation of nature and with human understanding, can and do always further the truth! As far as this book and its contents are concerned, however, I am not in need of a patron, since the book is not mine but from the pen of Theophrastus Paracelsus, upon whom the grace of God bestowed the knowledge of truth and medicine to conquer the evil in some persons, whether it be of the devil or his adherents. Truth, even though it be hit by envy, persecution, and hatred, is always desirable, and this has been proved by thorough works. But woe to those who consciously suppress the truth. As the first doctor to graduate from a university and take up the wholesome and honest doctrines of Theophrastus and publicly defend them, I have studied them and, by the grace of God, absorbed them and imparted them to those who need them, to all those who are in need of good will, favor, friendship, and kind assistance.

Therefore I am dedicating this very short book to Your Eminence; its contents clear and true are far more valuable than the writings of all those who had ever written on medicine before the kind and merciful Lord brought Theophrastus into this world. I dedicate it to Your Eminence that as a sponsor of the just Christian arts you may protect me and defend me as much as possible against the hypocritic and evil tongues, that the concern which I feel for your welfare and that of all Christians may be revealed and disclosed, and that all may know my reasons for abandoning the Galenic school. There has been much talk to the effect that I have been instructed in Galen's school and acquired not a small portion of my earthly belongings through it, and that I am a renegade and am ungrateful to my teachers. First of all, I must give an honest explanation of my actions, so that the truth may come to light now, during the lifetime of those whom I am going to mention.

At Basel in 1556 I was suffering from tertian fever (exquisita tertiana) and I called in two most learned and honest men, Dr. Oswald Beer and Dr. Johann Huber, who at that time were most famous physicians. However, following this tertian fever, I soon developed daily fever, and then quartan, and finally tympanitis. I was tormented by those diseases and by all kinds of medication vainly administered to me for a period of fifty-four weeks. When I found no relief through the doctors—who like true Christians had certainly given me their best advice and meant well—nor through my own knowledge or that of pagan writers, a dear friend, Cyriacus Legher, a physician—may the Lord bless him—brought me a prescription which had the following ingredients: *Spiritus vitrioli, Liquor serapini, Laudani,* etc. Legher insisted that this compound in solution was a remedy of proved worth against all kinds of dropsy. In my distress I accepted it despite the fact that I was taught to consider Paracelsus, the author of the prescription, an imposter, and I should have hesitated. As I was at the height of distress and pain I took the medicine as a last resort, using it as much as possible. By the grace of God I recovered from all pain and illness in thirty-four days.

At that time I was the personal physician of His Most Serene Highness, Prince Otto Heinrich, Count Palatine of the Rhine, Prince-Elect, and from my first association with him His Highness had admonished me to read the writings of Theophrastus. His Highness' gracious advice and the effects of Paracelsus' prescription induced me to consider the matter and to read his writings, and thus I became the clandestine disciple of Theophrastus, applying as much as I could the arcana and the medicines which he had advised. I was very successful with my patients, so that quite a number of envious persons said that I must be conjuring the devil since good effects were produced so quickly. As soon as I realized that I was on the right road, I purposefully and deliberately abandoned the old so-called " medicine " and turned to the natural medicine as the only one where comfort and help for all needs might be found; and I would like to see one Paracelsian remedy in all his books published through me which would not bring true results. I can say in praise that his remedies are based on truth. As fruits compliment the tree from which they grow and as righteous works praise their master, art has certainty in truth. (I am not speaking of those fatal diseases which bear death in themselves and which no physician can help.) Can any honest man begrudge me because I have been supporting the truth?

In telling how I adopted Theophrastus' teachings I must disclose two of the objections of those who want to censure me. These ungrateful men say that the writings of Paracelsus are a fraud and that although we, his disciples, are curing our patients of their diseases the cures are not permanent, etc. Such childish statements come not only from simple-minded persons but also from those who are considered wise but who reveal their lack of wisdom by such talk. It seems proof enough to me that I have cured palsies, lameness, the French disease, dropsies, epilepsies, gout, stones, and frenzies. Should I have promised them that their diseases would never return? Would such a promise be human or possible as long as they do not abstain from food and drink and from the elements, which contain all illnesses and their remedies? Has there ever been a physician who was able to protect man against wounds? Has there ever been a physician who could have made a person proof against fever, blows, falls, sadness, joy, anger, open wounds, internal or external injuries, and all the other things which may befall men, although that physician may often have cured a person of such ills? This criticism is no criticism at all. How can we effect permanent cures of severe illnesses? Far be it from us to indulge in such presumption and unchristian blasphemy, when all our help and prosperity come from Heaven.

If the Paracelsian medicines properly prepared are given in the right proportion every year, then they act not only through belief and confidence in the doctor but by their own proven effect.

There is a second criticism raised by foolish and contrary people against us—that we pretend to utilize metallic sulphur (salt of mercury), and that this is ridiculous. These critics maintain that a gold or silversmith could insure himself of a business through our writings, for the knowledge of how to gain oil from metals would be revealed to him. Another critic has said that he is fortunate in being a good linguist and yet he has never read of such formulae in either Greek, Hebrew, or Latin. Since he finds nothing of the kind described in other books, he draws the conclusion that all our teaching is useless and false. By their ignorance and lies about harmful results such erring men hope to incite people against us. To tell the truth, this is another reason why I need a good patron, which I am certain to have in the person of Your Eminence, for I have offered to prepare under your very eyes substantial amounts of quicksilver oil, and salt from such metals as copper, silver, and gold, and to do other things which are useful for the human body but impossible for the so-called physician to perform. Indeed, affectionate, dear sponsor, there are many such arguments and it is unpleasant to hear them all. Moreover, it seems fitting to let these petty quarrels rest and to speak about this litte book, its purpose and contents.

First, Paracelsus philosophizes on the origin and severity of the diseases depriving man of his reason; he mentions the sickness of the epileptic, all kinds of gout, insanity, St. Vitus' dance, loss of reason, melancholy and others. After this he gives excellent remedies, so that man can have no reason to complain that God has provided no help for terrible diseases. Although many diseases can be soothed by vitriol, it has seemed advisable to me to write on the method of bringing the arcana out of such minerals. I have extracted and dispensed them more than once and can give complete instructions on how to prepare and apply them. I have dealt particularly with sulphur,

as it is the greatest preservative I know. Until the present
time I have written little on either of these. Much has been
written in the book *De Vita Longa, Opere Chyrugico, etc.* on
the subject of metallic mercury.

I am well aware that many will reproach me, saying that I
should not so obviously throw pearls before swine, and that
the time is not ripe for such disclosures. To these people I
should like to point out that neither I nor any one of them has
a right to possess such a treasure, let alone to bury it, that such
knowledge should be turned to the advantage of ordinary
Christians and invalids to give them comfort, and all for the
praise and glory of The Almighty. Although true procedures
are being described here in a calm and clear way, the Lord
gives His gift only to those whom He chooses. The swine are
likely to perish before they bring their works to completion,
and birds who leave the nest too early soon regret it. Thus,
he who endeavors to seek the glory of the Lord and the welfare of his neighbor and not his own, deeming it righteous to
give and distribute with joy and a free hand, will certainly
find entertainment, pleasure, and earthly nourishment. For
who would have to suffer privation and beg for bread, if he
put his faith in love of God and of his neighbors? How could
a man not find earthly nourishment if, with the sweat of his
brow, he searched for the gifts of the Lord and, having found
them, applied them for the betterment of poor invalids?

Therefore, I wish to warn those who are cooking a slovenly
mess from Paracelsus' marvelous prescriptions for concentrated
diasulphur or diastybin; I also advise those who dare con-
found his and Galen's writings, which have less in common
than oil and water, to desist. Those who by robbery and theft
pretend that the salutary doctrines of Paracelsus are their
own are not worthy of being admonished, and I leave them
without warning. Almighty God will certainly put them to
shame and teach them to apply *aquas phlegetonticas* and other
malign corrosive " remedies " such as " Celestes " and balsams,
and with all their stoechomantic, frivolous arts send them to

Hechelberg through Charon. They should keep in mind that the Lord did not create the physician to be a thief or a murderer.

I have said all this in an effort to leave transmutations aside, and I hope that in doing so I have not incited envy, anger, or animosity against me on the part of intelligent persons. I hope that I shall win praise and gratitude, since those secrets are such evidence of the miraculous alchemical and medical works of God that one can never be grateful enough for the diligence, labor, experience, and love which Theophrastus had for his brethren, and for The Almighty who bestowed such grace on him and enlightened his eyes. Even if I undertook to write large volumes on such highly important scientific matters, I would be unable to point out their powers and advantages sufficiently. I have discovered myself that the medications, if prepared and separated from impurities and administered correctly, give such immediate, positive results that they are regarded as miraculous. I have received numerous messages from most learned and pious men who are physicians appointed to the Court and to the principalities of His Imperial Roman Majesty, our Most Gracious Sovereign, and also to the princes and towns of the Holy Roman Empire, that I may expect to receive acknowledgment and not slander from other doctors.

Now I commit myself to Your Eminence and both of us to Everlasting Almighty God and His mercy.

Dated, Basel, 1567, the day of Adam and Eve.

ADAM VON BODENSTEIN,
Doctor of Philosophy and Medicine.

Preface by Paracelsus

In nature there are not only diseases which afflict our body and our health, but many others which deprive us of sound reason, and these are the most serious. While speaking about the natural diseases and observing to what extent and how seriously they afflict various parts of our body, we must not forget to explain the origin of the diseases which deprive man of reason, as we know from experience that they develop out of man's disposition. The present-day clergy of Europe attribute such diseases to ghostly beings and threefold spirits; we are not inclined to believe them. For nature proves that such statements by earthly gods are quite incorrect and, as we shall explain in these chapters, that nature is the sole origin of diseases. In describing these diseases we find five kinds of loss of human reason, and as the first of them we shall speak of epilepsy, etc.

THE WRITINGS OF THEOPHRASTUS PARACELSUS ON DISEASES THAT DEPRIVE MAN OF HEALTH AND REASON

CHAPTER ONE

On the Falling Sickness

There are five kinds of epilepsy that deprive man of his reason. Each of them may cause falling sickness. One is in the brain, the second in the liver, the third in the heart, the fourth in the intestines, and the fifth in the limbs. The first three are more likely to prove fatal than the last two. The first affect the noblest parts of the body and so often lead to the death of the person who has been afflicted.

Falling sickness does not affect one sex only, but both. No one is too healthy or too ill, too young or too old for it. It does not, however, impair the health of the body, nor does it crush the natural being, for that is not where the root lies, as will be explained later.

This disease causes great weakness and death because its severity leaves traces behind, by which death is caused without impairing the structure of the person. As has been mentioned above, there are five kinds of epileptic disease. These five kinds of epilepsy have the same cause and the same symptoms, but they do not come from the same substance. They cause the same kind of swoon, although the substance and the paroxysm are different: in some cases the paroxysm is accompanied by much foam, in some by much water; in other cases the patients roll their eyes; some have twists and contortions; some are stiff; some scream; some are quiet; some are thrown against the ground as if by force; some sit down gently. We must pay close attention to these symptoms, for there is a great difference between those patients who feel the attack before the fall and then faint, and those who do not feel it before the fall and the fainting-spell. Cause and origin are not to be found in the agitation but in the organ affected by the agitation and by the attack; the brain is affected in the same way as the intestines, but in the first case the illness is more dangerous, since intellect and memory are located in the region of the brain.

Now we shall speak of the substance in which the disease is rooted. The disease exists not only in man but in all living creatures; the latter fall down in paroxysms in the same way as man. Some species of animals suffer from the disease by heredity and, being thus pervaded with it, none of their kind is without it, as can be seen in the squirrel and the lion, which become ill without cause. There are also many kinds of birds that suffer from it. Plants, too, which do not have any sensations, show this disease—not by falling down, for they have no sensitive life, but by having paroxysms just as sensitive beings do. Trees become split and torn, as we have mentioned in *De Infirmitatibus Arborum et Crescentium*; they may even dry up for some time, and then, after the paroxysm, either recover and grow again, or die. In everything that lives this disease is possible; the paroxysms differ according to whether the creature is sensitive or not. As to the real reason why neither sensitive nor insensitive creatures perish in the paroxysms we may say this: no living creature can exist without the *spiritus vitae,* which is the living strength of all things, as we have said in *De Spiritu Vitae*—nor is there any life [and that includes any maintenance of the *spiritus vitae*] possible without nourishment. Since everything that is alive may be affected by this disease, the illness may also be in certain food. When such food is mixed with the *spiritus vitae,* the trouble is brought about. This is one of the causes of the disease. The boiling of vapors in the *spiritus vitae* (but not in other humors and qualities which are also in the body) is another cause. If the *spiritus vitae* is shifted from its right disposition, it boils and effervesces, and this happens so quickly that memory and reason are destroyed. This boiling may be compared with the approach of an earthquake, which makes the whole earth tremble. Earthquakes and falling sickness have the same causes. Falling sickness is not a disease coming from nature in its regular course, or from ill health of the organism or destruction of the humors, but solely from the same causes as earthquakes, for the motion of the earth is also the motion of man and is experienced by all which grows on the earth. In our *Philosophia* we discuss the cause of the earthquake, which is the same cause as that of falling sickness in the body. If the living spirit boils due to faults in its properties, it produces vapors which make the whole body tremble. For if the living spirit lies down and suffers, the whole body lies down and suffers, not in bodily ill health but in insanity. Reason resides in the living spirit and is in the process of being destroyed until its destruction is complete, when vapor and brewing are no longer possible. Trembling, falling, foaming, and spasms of the limbs are caused because the living spirits of the patients are ill, and therefore the patients too are suffering.

The distinctions between the five different kinds of falling sickness are certainly clear. The first, in the head, can be explained as follows: the *spiritus vitae* boils in the head only, and there the illness starts; if the *spiritus vitae* causes vapor and corruption there, unconsciousness and insensitiveness result and carry away all reason. Then all limbs are weakened, because the root of the human being is suffering. Just as the earthquake takes place in one part but shakes everything within its reach, the disease is in the head only but afflicts all that it touches. It is the same with the liver: brewing and vapor creep into the *spiritus vitae* in the liver, and before the *spiritus vitae* can be felt it has reached the head and the whole body like a wind blowing along the roads. A piece of sodium, small though it may be, when it falls into vinegar makes everything boil; the action of the *spiritus vitae* is just as quick. The *spiritus vitae* is in the heart too. It boils and foams in the chamber of the heart, but, just like that kind of earthquake which does not make the soil barren, it is unable to cause poisoning unless the upper parts be broken by the shaking and trembling. Just like an earthquake wrecking a house, this breaking-up, however, is not due to the poison of the *spiritus vitae* but to its motion.

The same is true if the *spiritus vitae* has its roots in the intestines or in the abdomen. For the disease has a root from which it grows just as the herb has a seed which falls off, dies, and grows again. There is a root in the *spiritus vitae* as well, which brings forth a growth which causes such a disease. This root may be found in the whole body, wherever it may have been planted: if in the head, then it grows from the head—if in the liver, then the sickness grows from the liver—if in the chamber of the heart, then it starts from the heart. It is the same with the abdomen, where the *spiritus vitae* may be rooted in the intestines or in some other part. It remains where it is rooted and does not appear in any other part, because it has only one root. Sometimes the seat is in the limbs, the hands or feet, and will be felt in these parts in the beginning, from which it will rise up into the whole body by means of the *spiritus vitae*. It sometimes happens that this root is planted in the body but is unable to stay fixed because it is not very substantial or material. Consequently, it pervades the body and attacks reason.

Now that we have mentioned the location and causes we shall discuss the way in which the *spiritus vitae* may cause poisoning, so that sometimes there are paroxysms and sometimes there are none, and so that there is no sickness between the paroxysms. The disease begins in the mother's womb where it takes root; it is implanted in the children and grows with them. Weakness of the seed, slovenly habits and excesses cause the *spiritus vitae* in the child to be not as perfect as it should be and also make the ill health of the *spiritus vitae* a heritage. If a child is thus burdened and the *spiritus vitae* remains unimpaired, the latter will fight the disease until it reaches one place from which it can be expelled in the easiest way. The disease is not always immediately manifest, because the root is not always strong enough or large enough to show its noxious quality, but it does grow and does become stronger so that seventy years later it may be recognized. There are still other reasons for this: the fact that the nature and health of the body, and of other things that are in man, are in good condition and good order. Sometimes the disease appears after a shock, but the shock is not the cause of the disease, which had been lying, together with the root, and was merely inflamed by the shock, which causes the *spiritus vitae* to swell and boil. Joy may cause the disease too, by modifying that *spiritus vitae* which has an inclination toward disease. There are other incidents which may cause it, and in the following instance we see one reason why the disease does not remain and cause the trembling constantly: movements occur in the field of the influence which impair, weaken, diminish, or increase the *spiritus vitae,* and it is manifest after the root has grown, increasing in accordance with its strength and disposition.

In the other methods we shall describe what else should be known and remembered about this illness. In them we say that the disease cannot be healed in the root, but that it is quite possible to stop the growth of the root, just as the growth of a tree which has been felled and remains with its roots in soil but not in nature is stopped, as will be told hereafter.

CHAPTER TWO

On Mania

Now we must speak about mania, which is a transformation of reason and not of the senses. Perception is forced upon the senses, and there is no judgment at all. Mania has the following symptoms: frantic behavior, unreasonableness, constant restlessness, and mischievousness. It may be recognized by the fact that it subsides by itself and reason returns; mania may disappear and recur several times, or it may never recur. Some patients suffer from it depending on the phases of the moon, others after some external accident. Thus we have two kinds of mania and, likewise, two causes: one coming from the healthy body and one from other diseases. Both kinds are kindled in the same matter and in the same way, as follows: the matter from which mania grows is a distilled humor which enters the head, after being mixed below the diaphragm, above

the intestines; another mixing takes place between the diaphragm and throat, going directly to the head. There are two
matters in the body that by being distilled are able to cause
mania. In the extremities distillations may take place too, moving upward through canals and pores. Only from these three
can mania come.

We must distinguish between the manias, for each distillation has a special origin and effect, one being more dangerous
than another. Thus the distillation coming from below the
diaphragm makes the patients almost mad and senseless. They
fall down quickly, do not want to eat, vomit a great deal, and
suffer from diarrhea; they speak to themselves, do not take
notice of other people or of their surroundings. This is because
this mania is produced from excrements, which dissolve and
are sublimated and lead to the symptoms described above. The
mania which grows above the diaphragm is very grave and is
accompanied by much pressure around the heart and pain in
the chest, because the matter is distilled there too and rises and
therefore causes such pressure. The mania which comes from
the limbs makes the patients gay and joyful and also quite
wild, because nothing is felt in the inner organs. They become
gay and wild and frantic. Although there are other symptoms
by which mania can be recognized, there is no need to describe
them. There are two manias: the first, in which the distillation penetrates the head and remains there like a vapor, will
remain as long as there is such a vapor; the second is sublimated
in the head, not by distillation but by coagulation, and it
remains there undissolved.

Therefore, for the cure we have to distinguish two kinds of
mania: one caused by distillation, the other, which has to be
dissolved and consumed, caused by sublimation. Although
there are two kinds of " elevation " of manias into the head,
they rise from one basis. There are three kinds of mania but
there are many more ways in which it can appear: some distillations and sublimations take place exclusively in the blood
of the veins of the extremities, some only in the tendons and
nerves of the limbs. Therefore, we have to distinguish between

two paths of mania in the extremities: one through the blood, and the other through the nerves and tendons. Sometimes the mania which comes from the extremities is not in the whole limb but only in the blood or in the nerves and tendons, and is distilled upward through them. Such distillation comes first to the heart and then to the head, because the veins and arms lead the distillation and sublimation to their origin, from which the mania rises. The mania above the diaphragm may rise from the stomach and, after having dissolved the phlegm in it or the matter lying in it, go up into the head. The liver and the lungs may also cause such distillation and sublimation. It is the same with the intestines and with the kidneys: some distillation or sublimation comes from the salt, some from the *spiritus distillatus,* some from the mucus in the intestines and from other things contained therein. Thus, each can give mania a separate name.

Now we shall relate what causes mania, how it can be dissolved and how it becomes sublimated and distilled, and we shall speak of its origin. It is caused by great heat in two ways. One is hot like fire. If the matter from which mania comes vibrates into chalk which dissolves in a very corrosive water, this corrosive solution sets the *humor vitae* afire, and this causes a separation of the rough from the fine. The fine parts are so sharp that they go up as far as they can, while the rough remain below. Such vibration is caused by great heat in these places, which rises and burns until it makes these fine substances. Just as there are various kinds of fire there are also various kinds of solutions, differing in the blood from those in other parts in such a way that the *spiritus salis,* which is in the blood, mixes and becomes one with the *spiritus vitae.* Now we can understand why the distilled humors which rise in the heat of the other limbs are so fine and sharp and cause insanity on touching the particles of the brain. For this reason, if the *spiritus* is distilled only in the blood it will come out if the veins are correctly bled. If, however, it is not distilled in the blood alone but in the whole head, no bleeding will help, as we mention later in speaking of the cure. It often happens that by opening fingers and toes mania can be made to disappear, because coolness and air are thus admitted which prevent the matter from getting warm and rising. If this does not always or completely help, it is because the matter cannot be cooled sufficiently to let out the distillation. If mania is in the veins, bleeding will not help either. In that case one has first of all to find the origin and start at that end, as we shall indicate later in discussing the cure. Further, mania which sublimates and coagulates in the head makes worms grow there, due to the putrefaction of the light sublimation; sometimes there is an ulcer if too much hard substance is congealed, sometimes there is pus all through the head. This causes much gnawing and a stinging pain in the head, with much nasal mucus. Also, it often happens that such solutions in the head are consumed again, except for a tiny drop which by itself may cause mania if it lies in a significant place. If the matter does not dissolve, however, it remains coagulated and burns and gnaws—as when salt is put into a wound—and it deprives man of reason.

The place where mania is originally located, which we can indicate and understand only in a general way, cannot be absolutely determined. The sublimation may start in the stomach alone, even though the stomach is small, and in not the whole stomach, but only one part. But if the whole stomach is given a medicament and the medicament works, mania is removed too. It will then be noticed that this cannot be done in the case of the veins, because somtimes the mania rises in one vein only and sometimes in all. The same is true of the nerves and tendons. As far as the liver, lungs and spleen are concerned, the whole organ should be given the cure. We must also bear in mind that the ancients, our ancestors, said that there were four constitutions which caused such mania. We cannot understand why melancholy should be a cause. If melancholy were a cause the melancholic part would be the only one to suffer; the same would be true if yellow bile were the only cause, and this could be cured; and it is the same with blood and phlegm. But according to our experience this is not so. For if mania starts in three places of the body it cannot come from one source, for it would then seem not properly distributed. Also, if mania came from constitutions, as has been proposed, no distillation or sublimation from the external limbs to the head could take place. For none of the constitutions has the disposition to become fine enough to numb the head. Mania begins in those limbs where no complexion [constitution] rules. If this is the case it cannot be derived from the constitution, as we have stated in *De Complexionibus*. Insanity may come from an organ where melancholy rules, such as the spleen—not, however, because of the melancholy, but because of the *spiritus vitae* which in the way described by us separates from melancholy and rises. Therefore, mania comes only from the *spiritus vitae* and in consequence of sublimation and distillation in the head, as has been said. Also, the insanity coming from the gall is not due to the yellow bile therein. As we must go on to speak of other things, we shall state our conclusion that the *spiritus vitae* and all matters from which mania comes are of one nature, quality, complexion, and are alike in one being; one is not hot, another humid, another cold, another dry, but all equally bring mania, because the matter is so fine and sharp that it clouds the brain and through it the cells of reason. Thus man becomes maniacal.

Now we have to consider the period of insanity and also the gestures which might perhaps indicate that the disease came from the influence or the qualities, which is not the case. For it is quite possible that it increases and decreases with the influence, but it does not come from the increase or diminishing of the influence. As we state in *De Influentiis Humanis,* the influence does not rule our body materially or physically, but our innate influence does. Thus we state in *De Mania* that it comes from the human influence which is both in us and those above, of which we do not speak here. From the gestures (which do not come from qualities) one might be led to believe that the maniac who mumbles is a melancholic, which is not the case, or that one who wants to fight and kick is choleric, which is not true either. The gesture and behavior come from the fact that such a melancholic, who by his nature has been a natural melancholic, becomes a maniac. The maniacal substance irritates his former behavior and the gestures inherent in his nature and makes them apparent. For mania incites secret gestures and qualities which men have hidden in themselves.

There is also a mania which does not indicate the nature of man; on the contrary, his nature is against it and rebels. Therefore, watch how the nature of man displays itself, for it often happens that the constitution of man becomes manic and he attempts to expel the mania. In such cases the natural qualities of mania are also seen. It follows that, as we stated in the beginning, mania does not result from the qualities, as it appears to. Those manias which develop from sublimation of *spiritus vitae* and the like are not due to melancholy, but merely show the same symptoms as melancholia, etc.

CHAPTER THREE

On the Origin of Truly Insane People

In previous chapters I have spoken of the deprivation of reason, but it is not the case that patients remain without reason until they die. They suffer from attacks time and again, so that they lose their reason and then regain it, as we have indicated. Now we shall speak of those who are permanently insane and of insane body, rather than of those subject to recurring attacks. The period is irregular, once long, then short, in correspondence with the stars; it does not always occur, behave, and stay in the same way, but is irregular in accordance with the course of the stars.

There are four kinds of insane people: *Lunatici, Insani, Vesani* and *Melancholici. Lunatici* are those who get the disease from the moon and react according to it. *Insani* are those who have been suffering from it since birth and have brought it from the womb as a family heritage. *Vesani* are those who have been poisoned and contaminated by food and drink, from which they lose reason and sense. *Melancholici* are those who by their nature lose their reason and turn insane. We must, however, note that apart from these four kinds there is another kind: these are the *Obsessi* who are obsessed by the devil; the various ways in which this happens are treated by us in *De Spiritibus*. But here we deal with those who are insane by nature, and sufferers of these four kinds cannot become obsessed by the devil and his company, as many people say; for the devil and his crew do not enter an insane body which is not being ruled by the entire reason according to its quality. Therefore, the devil does not enter those four kinds of insane people for, due to causes that will be explained later, they have no power of reason. While they are out of their senses they are possessed by, but safe against, the devils and *Vatis,* as we have said. The reason for this will be given in the proposed chapter *De Spiritibus et Vatis.* Although there are four kinds of insanity, each with a separate origin and derivation, they all end in depriving man of his reason, not in the same way, as we announce in the first four chapters, but without any other disease, so that there is no pain, as in epilepsy, in *mania,* in *chorea lasciva,* in *suffocationis intellectus;* but these people always live in madness and, in the cases in which it is apparent that there is going to be another sick day, increasing insanity. When reason announces such sick days, death is not far, because the origin of the disease is so violent that it hurts the *spiritus vitae* and poisons it, and death results.

Now we shall take up the lunatics and the reason for their disease, so that the discussion of them in the second chapter on *Methods of Cure* will be understandable. The stars have the power to hurt and weaken our body and to influence health and illness. They do not fall into us materially or substantially, but influence reason invisibly and insensibly, like a magnet attracting iron, or a scarab dust, or asphalt fibers and wood. Such power of attraction is possessed by the moon, which tears reason out of man's head by depriving him of humors and cerebral virtues. The moon does not enter us and work in us, as has been affirmed, for no star has the power to possess us, as many state falsely; but we must believe that it takes reason from us as the stars do by virtue of their power of attraction. Just so the sun takes the humidity out of the earth, not by entering the earth and driving out the water as if it had been poured into fire, but by attracting it. Not only the moon acts thus; many planets deprive the organs of the whole body of their humors, as we discuss not here but in *De Astris et Superioribus.* Thus many people are deprived of their senses only by the moon's attraction which takes out the cerebral humor so that the whole head rages without reason. Those lunatics are ruled by the moon. The power of attraction is at its height during the full moon, and therefore it attracts most strongly and the lunatics suffer most then. It is somewhat the same with the new moon, because the newly conceived moon has new virtues which cause greater or lesser insanity; it is not that the moon because of its weakness weakens the limbs, as if all our power lay in the moon, as we shall point out in *De Firmamento.* But owing to the unusual nature of the moon there are various degrees of attraction. It so happens that the humor is drawn out of the brain when the moon is smallest. The explanation is that the new moon does not draw out the same humor as the old one: the new moon draws out the body's humor less than the old moon, which is rough and hard and therefore draws out the rough and hard humor. Just so, a young fire which has no great burning capacity does not drive out oils and hard substance, while a great fire takes out hard *and* soft. When the sun rises it takes away only the moisture which is not heavy but light, while at noon the heavy waters are taken. The removing of moisture is more noxious to the earth and its strength than the removal of the rough, great humidity, due to reasons given by us in *On Dew.* The sun at noon absorbs dew and fine particles, rough particles, and vapors. But this is not the case with the moon; it is not in its nature and quality. When it first comes out it takes the fine humor from the brain so that the full moon is compelled to take the rough, for nothing fine is left; under the waning moon the humor grows like dew at night, and when the moon is on the increase it is driven out

again. Many people are more insane and frantic during the
waning moon than during the waxing moon, because the full
moon has already taken the humor from them and they suffer
more acutely the more the moon wanes. Just so the splitting
of the earth from lack of humor cannot take place in the fore-
noon but only in the afternoon after the sun at noon has taken
away the humidity. It may happen with man that his insanity
shows only at the end of the moon's period. There are various
reasons why the powers of nature struggle increasingly until
they can do no more. While the moon exerts its attraction, as
long as there is humor in the head it will draw it out. This is
the main reason for its drawing out and attracting most at the
end of its period, and this explains why the end is most trouble-
some for those who begin the sick days of their nature during
the moon.

In the same way we can explain those people who have
received insanity from the mother's womb as a heritage, such
as a family which is insane or a child who has been born insane :
the seed and its function may be defective, or it may be inherited
from the part of father or mother. The first reason is that the
sperm in itself and in the operation may be lacking in the
power of matter which makes and builds the brain. If the
matter of a limb dissolves, its proper shape and nature dissolve
too, as we have said in *De Generatione Hominis* in Volume
One of our *Philosophy*. If there is such a deficiency in the
sperm, reason is not complete since matter is not complete ;
therefore, there cannot be completeness of reason in the child.
There is another cause for insanity in people, which is produced
during the development ; if, at its height, the power of attraction
of the moon interferes with the generation and conception of
the child, the moon can take away reason, so that complete
sanity can never be restored. Such deprivation does not happen
every day, although the moon rises every day, because the
matter and the moon must be in accord with each other, and
not every cerebral humor can be attracted by the moon, only
the one adjusted to it, as we have stated in *De Generatione
Humana*. If it does happen, the deprivation of the senses takes

place too. The moon can take away man's reason at birth; when, however, this is hereditary, the circumstance is such that if there is insanity in the brain, the child's mother also has some deficiency in her brain, for the brain of the parents is continued in the brain of the son, as we have said in *De Generatione Hominis,* because the nature and qualities of the one originate in those of the other. This does not always happen, because the sperma become mixed, and either the man or the woman may or may not be insane, and the child may follow the insanity or take after the one who has the greater influence. It may even happen that if both are insane they still would give birth to a healthy child. This is due to the power of nature, which drives out the impediments and adversities.

Now we must also speak about *Vesani* who become insane from eating and drinking. It often happens that food offered by whores causes deprivation of the senses, and this in many ways. Such insanity may lead to love, so that the *Vesani* put all their being into the whores; some are bent on war only, and therefore they have to do with war only, and there is no sanity; some climb and run; some act in various ways which we do not wish to enumerate and shall not describe, but which should be remembered. Let us not be amazed that it should be possible for food to cause this, for it is possible; and much less should we be astonished at the effects they produce. And why? Because food and drink have affected them greatly. Now we shall describe the fourfold insanity that comes from food and drink; we shall have more to say about it in the chapter *On Treatment.*

First of all, there are those that have so eaten and drunk that they have to and are compelled to love a woman. We shall leave out a few points here. If a person offers something to eat to another, whether man or woman, unbreakable, eternal love is the result; for this reason some servants give food to their masters so as to flatter them and to make love spring up in them, with the result that servants are above the masters, as we shall mention in *De Republica.* Further, unreasonable animals, dogs and others, can be brought to such love for those who offer it to them. It is natural that women should give such things to men, so that the latter should think only of making love to them; and they have no reason, and their melancholy is directed toward those women who have offered it. We shall leave it at that, because of other proposed topics.

The others who are bent only on war have been given some food intended to make them insane, and if they are choleric by nature they think only of war; their insane behavior has been given them through the food. We must also speak about the melancholic and phlegmatic ones, who exhibit their nature and constitution in like behavior.

The third, who jump up and run around all the time, have received their insanity from eating the kind of thing that gives them an urge to mount and climb; this comes from the nature of this food and not from that of man. If it were up to us to write about it we would reveal everything here, but some things must remain unmentioned as there exists a great body of philosophy and contemplation on the nature of this insanity, and that must suffice.

The same is true of the fourth kind, which we shall not describe at all; it may deprive man of his reason in the way described above. Some incantations are able to do the same; we shall not describe that here but put it under *De Influentiis*. After the first accounts in this chapter we shall speak about melancholy persons, of whom there are four kinds. If such complexions deprive man of his reason, it is due to their driving the *spiritus vitae* up toward the brain so that there is too much of it there. We shall not speak of this here but leave it to those who write on philosophy.

On St. Vitus' Dance

We do not wish to admit, in this chapter, that the saints can cause a plague or the diseases which eventually are named after them. In our opinion, such diseases have nothing to do with

the works of the saints. There are many who connect great
sufferings with saints and attribute their sufferings to God
rather than nature. This is idle talk. We dislike talk behind
which there is no proof but mere belief. Belief is an inhuman
thing and the gods do not think much of it. As this disease
is known under the name of a saint, we do not intend to change
the name; it should, however, rather be called *chorea lasciva,*
for reasons which will be given. We discard unproven words
which speak of God without knowing Him and digress from
His given path by which He may be recognized. Thus the
cause of the disease, *chorea lasciva,* is a mere opinion and idea
assumed by the imagination, affecting those who believe in such
a thing. This opinion and idea are the origin of the disease
both in children and adults. In children the cause is also an
imagined idea, based not on thinking but on perceiving, because
they have heard or seen something. The reason is this: their
sight and hearing are so strong that unconsciously they have
fantasies about what they have seen or heard. And in such
fantasies their reason is taken and perverted into the shape
imagined. Moreover, in adults who do not imagine the dance
but hear and see it, hearing and sight become stronger than
reason.

Even people who do not have such sight and hearing are
afflicted with the dance; they are overcome by this kind of dancing and joy. The cause is in the laughing veins which comprehend their spirit in such a subtle way that they are tickled into
dancing and joy. There are two causes for this disease, for this
dance: a natural one from the laughing veins, and an incidental
one from the imagination. The natural origin of the dance is
this: everyone has laughing veins; if these veins are opened
and bled, the person begins to laugh and cannot stop while
he is bleeding. If the bleeding continues he will laugh until
he dies. Such veins are the cause and origin of this disease,
and this is what happens: the veins remain complete and
uninjured unless the *spiritus* in them on which they live and
feed should be modified and separated and thus digress from
its course and order; then it jumps and makes the blood rage;

this causes a ticklish feeling, then laughter, which in turn makes the spirit in the veins stir more and more, for the veins lie at ticklish ends and spots and it is in their disposition and nature to make one laugh. The reason for the *spiritus vitae's* moving and breaking is the fact that it is a fine *spiritus*. The life of those veins can no longer be regarded as natural. Just as brandy left to itself becomes sharper, finer, and lighter from the warmth in the wineskin, the *spiritus vitae* in the veins becomes fine and sharp from natural warmth, and as a result the blood changes so that it has the same quality as ordinary wine after its substance has been changed by mixture with brandy.

We must consider the origin of the *spiritus vitae* in those veins, what makes it ailing or disposed to become so fine and sharp—for it is not in its natural disposition to do so. The cause that incites it to become so fine and to move the blood into the disease is this: the blood contains a salt which has a natural disposition to turn sour, sharp, bitter, or sweet, depending on circumstances. The material and physical parts of the body are inclined to change, just as wood putrefies or turns into ashes or coal by itself or by accident. The salt is changed in the same way. The change makes the *spiritus vitae* heated and frantic, not because it is hurt—for it has no body—but because it is resting in an unnatural place; just as camphor cannot remain unchanged in a dirty place, so *salniter* cannot stand its counterpart in fire. Another kind of dance may result from stimulation which originates from vision and hearing, in this way: joy in man comes from the heart; vision and hearing are things that go to the heart. If I hear someone whistling, something I like by nature, I feel joy in my heart. This joy is twofold: I have a feeling of pleasure as is my natural disposition, and, besides, I have the image in my senses as if I see the person whistling before me. And while I am pressing that whistling into my thoughts it gives me pleasure, and joy prevails in my heart; as it stands before me it impresses itself upon me while all other qualities, blood, and dispositions are driven from me, so that they are suppressed and have no further effect. This is followed by the deprivation of senses, but not of reason. If my power of reason is taken from me and, due to my imagination, I act in the same way as I would if I had noticed or watched the whistling, my lack of will is the cause of my disease. And it is natural that such lack of will, in which a person indulges with joy and with all his heart, causes such imaginations, as we have described in *De Imaginationibus*. This obviously is one of the reasons why whores and scoundrels who take pleasure in guitar and lute playing, who satisfy all voluptuousness, bodily pleasure, imagination and fancy, never escape but become ill in such a way that they jump and dance, thus applying what has been their occupation. This dance which we find in whores does not come from nature nor have a natural cause in the laughing veins, such as we have described in the beginning of this chapter, but it comes from recklessness and disgraceful living in which there is neither reason nor sense. It is for these reasons that they behave so disgracefully and unreasonably. But there are also many who do not lead such a disgraceful, inhuman life, who do not think of it, and therefore would neither want nor enjoy dancing; they dance without thinking. The cause is the laughing veins of which we have spoken. The dance also comes with laughing, howling, screaming, or jumping—sometimes only with laughing and walking. The cause is again in the laughing veins, with only the difference that the *spiritus vitae* does not change—the blood is not poisoned but is inflamed, and it subsides after a while. Now we can also understand in what way much of the joy comes from the heart: the heart feels a salted *spiritus* which stimulates it to laugh, and the heart is powerfully inflamed by the *spiritus vitae* in the laughing veins. For the same reasons joy may befall the heart just as illness may the spleen; both are possible. It may also happen that such joy in the *spiritus vitae* rises into the head, takes possession of it and then acts accordingly. In such cases the patients have no great urge toward, or pleasure from, dancing; they are willing to perform what you want them to do, and this without sense or reason. The other dancers, however, are forced by their illness to dance, jump and scream with many gestures. Now we can understand why they have to jump and dance, etc., in the way we described in *On Imagination and Fancy*. The natural cause is in the laughing veins which have a quality which makes a person jump, laugh, etc., if he gets tickled at the ticklish ends; and the same tickling happens in the veins, which contain even more laughing, tickling, and jumping, for their life is inflamed and is boiling within them. The dance of this disease is also a modification of reason, because joy takes the lead and changes all other qualities. It does not poison, or deprive the patient of, memory, but memory does become disturbed, and it is suppressed and hindered by the disease.

We shall have to speak about the duration of this disease: the disease appears for a period which is sometimes long, sometimes short, depending on the quantity of the matter and on the disposition, by which it announces itself, grows, and rises. Now we shall come to a close concerning this disease of the dance. We must not underrate the power of the gods who are able to punish a voluptuous person with such disease. The nature and origin of the disease cannot be removed or changed, for such disease is an actuality. In *De Plaga Dei* we shall say more about this illness, and at the end of the chapter *De Cura Chorea Lascivae* we shall give further details about the dance.

On the Origin of Suffocatio Intellectus

There are also several kinds of deprivation of reason, coming from natural diseases, which make man insane. There are three of them: one that happens only to those who have worms in their intestines, a second that comes only to women from the womb, a third that befalls both man and woman from lying or sitting or through overeating and overdrinking. There is another that comes from sleep and subsides by sleep. Therefore

we have two types: one beginning in a state of consciousness—
so that the patient know what is happening to him, and the
other in an unconscious state—which comes from outside and
makes the patient ill without anyone's noticing it. One has to
know more about this than about any other illness leading to
a deprivation of senses, because the patient's life may be lost
if the doctor has not insight and knowledge. It is not as if
reason carried death, but the disease which takes reason away
also destroys life. To those who have worms, as we mentioned
in *De Vermibus,* death comes with a convulsion; also, the life
of women can be squeezed out if the womb is contracted, and
many other things. Death may also be caused by lying and
sitting because the air is taken from the heart, as we have
stated in *De Introitu Mortis.* In this disease there are many
symptoms by which future good and bad, life or death may be
recognized. For the patients fall down in just the same way
as in epilepsy and have cramps which cause contraction and
extension; such gestures are equally distributed among the
patients and are the very sign of the disease. The deprivation of
reason does not always appear with such shock and trembling,
but sometimes with a quiet sleep, sometimes with a disturbance of reason, for there are many varieties of this suffocation
—and it can always rise and grow again. As to the worms, they
can grow and lie in many different places and are therefore
apt to cause many severe, strong convulsions. This is also true
of the other varieties of deprivation of reason.

We find this suffocation in two places: in the mother's
womb and in the belly. It can also be limited to the head, but
in this case it is not caused by nature but comes incidentally,
from blows or wounds which are its cause and motive, as we
shall explain later. First we have to understand how suffocations originate. There are two ways: one by smoke, in which
case a headache follows the convulsion because remainders
are left in the head and change its sensibility; a second by drawing the air toward the heart, in which case the patients suffer
from a sickness around the pit of the stomach if the latter contains some substance from which worms may grow or have

already grown. This substance, or the worms—if separated from the feces, or lying there and putrefying, or devouring each other, or sublimating the substance in the stomach, which then overflows—cause a smoke which rises and clouds the brain and also the sleeping veins, so that the patients fall asleep without knowing it and dream under heavy, great pressure and with difficulty. The entire reason may disappear, depending on how the sleep is being disturbed, as we say in *De Somno*. For such an injury can also injure the brain, from which the insensibility of sleep comes. It is the same with the worms which are in the stomach and in the opening of the stomach. The worms in the belly lie in the intestines; they do not rise into the head as the stomach worms do, but produce their convulsion without any smoke and vapor to impair the head. Their smoke and formation cannot rise into the head for many reasons. This smoke and shape with its vapor injures the surroundings of the heart, however, so that the air and nutrition which go to the heart are poisoned. And if the heart receives them it chokes in them, and death may occur from so much noxious substance. This cannot happen in the head, and after the effect of the substance is gone complete health follows; but the convulsions succeed each other. If the heart suffers all living spirit suffers with it. This causes cramps, trembling, shudders, and tantrums with many other symptoms, resulting in an insensibility in the whole body, so that there can be neither sense nor reason—to such an extent do blood and humors boil and rage in the body and against each other. It is as if sulphur and *sal nitre* were put afire together. Unless the noxious matter has been consumed, this insensibility does not stop or else it strikes at life itself.

We have also to speak about the womb. It produces, though in another way, the same signs and symptoms as those appearing when the womb's natural condition is changed into an unfavorable one, which change results in a contraction of the uterus and takes away all reason and sensibility. The cause is that the matter on which the womb is internally nourished and kept alive destroys itself, like wine turning into vinegar. If the womb neither feels nor has the proper substance, then the substance has lost its right nature and is cold. This coldness causes a tension of the skin and of the lining of the womb, contracting it into a kind of spasm, for it is an inherent quality and innate nature that all acid and cold of that kind give a sort of spasm or stinging to the whole body, with the exception of the flesh and bone. This causes coldness and sharp acid in the uterus, resulting in cramps and contraction, making the uterus like a log. The contraction, the tetanus and the spasm also force the other limbs into spasms and tetanus, for they become contaminated by the womb also. If such contraction takes place in the veins of the whole body, vapor and smoke come out of the womb to the organs around it, and if it touches the heart the convulsion is similar to epilepsy with all its symptoms, and no other organ but the heart is afflicted with it.

The third kind is another suffocation which happens in many and varying ways and which we cannot describe in all its details and varieties. One comes during sleep if one is lying on one's back. The water of the heart sinks into the lower chamber and the heart above remains bare. Some people believe that this causes a weakening of the heart, for if the heart lies in any other place than in the water it deprives the patient of his senses; it rages and takes all power and strength out of the limbs—out of the blood which is being driven there. This belief is not correct, for the blood cannot do any harm to the heart except in other illnesses where one has to lie on one's back, where it may have many causes. Some also believe that the blood retreats from the heart, so that it lies bare by this deprivation, and it seems as if the sleeper could not move hands or feet. This is not so. For there are so many other reasons why the blood cannot flow away from the heart. This flow would be very injurious to the heart. The real reason is that, while one is lying on one's back, the chamber is pressed flat, just like a bladder which is filled with water and can be squeezed. If the heart is bare, or touches the skin of the chamber, it causes pressure as if a heavy stone were lying on it. For there is much around the heart, the matter of which might fall on the center of the chamber, if one is lying on one's bed and pressure is exerted. It cannot happen when one is lying on the side or on the belly, for reasons discovered in many ways in experiments in anatomy. So we need not go into them any further.

There are still many other diseases in sleep which deprive man of reason and also make him frantic and wild, like falling-sickness. There are two causes: one is the same kind of pressure as in falling-sickness, the other is a pressure of the brain. This happens in the following way. If there is pressure on the heart so that the *spiritus vitae* grows hot, the whole body is inflamed and the pressure becomes so strong that it takes possession of reason and of heavy sleep, changing it into trembling and pain, as if it were epilepsy. It can also happen in the following way: if the disease returns to the belly again it does not leave the body. Why? Because the *spiritus vitae* is still frantic and raging. This is not felt in the morning after sleep, and after the convulsion the patient excretes a yellow water produced by the *spiritus vitae* during the frenzy. It may also happen that the *spiritus vitae* leaves all organs and lets them lie as if they were dead, and only in the heart, brain and liver is there any life; elsewhere there is none. The body seems then like a tree which has all its strength in its roots but seems dried up and rotten otherwise; yet strength may be restored to it. The *spiritus vitae* is also able to spread throughout the body from the center or from the root, after evaporation of the humors, and to call back life as before, as we shall say in the chapter on *Cure*. The same kind of pressure that we have been describing in the heart may also occur in the brain in so far as the sleeping veins can compress the brain and reason by their pressure—in the *pia mater*—just like a suffocation which compresses the uterus in which the disease is rooted.

This is the cause: sleep changes the whole head and presses the *spiritus vitae* into the brain until it evaporates; then it appears with all symptoms, spasms and trembling, as we have said about the heart and the uterus. Sleep can be injured still more easily than the heart, without loss of sanity. Intoxication can cause such suffocation too; this is to be understood in the

same way as the worms in the stomach. It does not come from
a matter in nature but in the following way. Drunkards have
a fine spirit which by its sharpness can injure sleep and sometimes the brain, as has been repeatedly reported. Intoxication
can also cause mucus in the stomach similar to the matter of
worms and having the same effect as worms. It can also happen
through drink and food which do not contain such fine spirit—
for example, thick sauce and water. The cause is an impurity
in the sauce and in the water, which have an inclination to such
disease. It is not in all thick sauces and waters, only in some:
water which has been poisoned by worms and their excrements,
for example, or vegetables on which poison has been poured,
or mosses on which there is such excrement as has been
described.

It happens quite often that a sudden convulsion comes during sleep in the form of a suffocation or something similar,
as has been reported. We do not write about it here, but leave
this to others, for what is possible during the day is also possible
during the night. It should be kept in mind that the course of
the stars is also able to cause or remove such diseases. It often
happens that, depending on their influence, such convulsions
may or may not occur, for the *spiritus vitae* takes ill and becomes
sad accordingly.

There are still other deprivations of reason, without fits
and frenzy, however. For example, one may sit and turn quite
limp and fall down because the matters of this disease, as we
have told, have not been able to inflame the *spiritus vitae* so
quickly. Reason may be lost by incidental blows and similar
things, as a wound on the head, which will cause fits of frenzy
several times a year. This happens because the same influence
rules the body and the stars, as has been explained in *De
Influentiis*. At the time when the influence reaches the head
it also touches that blow and this causes frenzy at certain times.
It may have other reasons too: the wound may be badly healed,
pus may have accumulated, inflaming and injuring the brain.
There may be still another reason: the wound may have healed
together without fault, but the brain may have suffered such an

injury by the blow that it can never heal again. It makes itself felt depending on time, weather, exercise, food and drink, and many other factors which need not be described as this will be reported in the chapter on *Cure*. Now we have said enough about suffocation.

The Second Method of Theophrastus
On Soothing and Curing Falling-Sickness

THE FIRST CHAPTER

Having described the origin of falling-sickness in the first part, we shall now undertake to explain its cure. Do not admire the power of the cure of this disease, for all diseases originating in the body have also a cure in medicine. Up till now it has been said that it is impossible to cure this disease because people did not know more about it or perhaps concealed something, as we shall later report. The basis upon which we found our medicine, and which has been given us by experience, is not to think back but to describe two kinds of medicine: one affecting its kind in a physical way, and one affecting its kind in a spiritual way. Thus we understand that there are two kinds of disease in all men: one material and one spiritual, which we explained in *De Principiis Sanitatis et Egritudinis*. Against material diseases material remedies should be applied; against spiritual diseases, spiritual remedies. Thus we say that falling-sickness is a spiritual disease and not a material one, so that no material medicine would be of any help. Therefore, the remedy against a spiritual disease is, and must be, spiritual. There are no spiritual remedies except those especially made for it, as we shall indicate later in our practice. This can be understood best in two ways: first, there are some material remedies which cure spiritual diseases, as is done by the following prescription in new and very recent cases; it is a material remedy and cures epilepsy:

Recipe for Epilepsy

Take *camphore, spodii, unicornu,* etc., *fiat pulvis*

This should be offered in a soft-boiled egg, also the powder. Same compound also soothes certain epilepsies. Now we have indicated the material medicine against this grave illness. Its effect simply consists in an obstruction of the canals through which the disease is rising. Its effectiveness is also based on great cold, which makes the matter coagulate and die, so that it becomes insensible, like a frozen limb. This is done in the following way: take *camphore,* etc. There are some who cure it by the specific properties of *paeonia, viscus quercinus.* Some make it insensible so that it cannot break out; this is done in the following way: take *opii mandragore,* etc. Such medicines are merely material ones. Therefore, they help in fresh cases only, where the disease has not yet got the upper hand.

Now we shall proceed to the spiritual medicines which cure epileptic patients; the remedies we have mentioned before neither help with certainty, nor help intrinsically in all cases. We shall not follow the writings which are opposed to us or those which do not allow us to proceed in such a way, nor shall we put faith in them; for experience teaches us more than our opponents have understood or ever will understand. Nor shall we specify our cure in analepsy, catalepsy and epilepsy separately, but we shall regard it as one falling-sickness and practice accordingly; we shall not withdraw what we said on the origin of falling-sickness; our prescription will conform to that description.

While we are describing the cure four points should be remembered:

1. The medicine should become spiritual and fine; it should be a coagulation *post spiritum.*
2. It should be administered with veins compressed.
3. It should be made specific with appropriate means.
4. It should be prepared from herbs which have a fine spirit and go through the entire body like an elixir or arcanum

or like a quintessence. There is no other way of conquering the disease. The remedy must have the same strength as the disease.

Now we shall enumerate the four points, what they include and what spiritual quality may be drawn from them to oppose the disease. First:

Coagulants:

Camphora, nenuphar, cristalli, spodium, sinapis, sperma rana-rum, unicornu, coralli, mumia.

Constrictiva:

Bursa pastoris, verbena, salix, coralli, hypericon, rosae.

Specifica:

Viscus quercinus, poeonia.

The subtlest spirits are these: *vitriol. ungar., vitri. cupertinum, vit. romanum.* If these four are in power, we find some others which have a marvelous effect if mixed into the spiritual spirits; these are: *opium optimum, mandragora, papaver, lolium, hyosciamus, uva versa.* We do not utilize them in the way in which they appear, but according to the use and nature of Archidox preparations. We shall never be tired of describing the virtues of this preparation, which we shall be glad to explain in the following prescription. There are also other medicines, like *medicamenta confortativa,* which by their excellent substance help nature to drive out falling-sickness. These are: *aurum potabile, oleum auri, quinta essentia auri, materia perlarum, corallorum solutio, magisterium antimonii, extractum sulphuris, mercurius reverberatus.* These medicines have a wonderful power. It is hardly believable that such things should be hidden in nature, to cure diseases which cannot be cured in any other way. Therefore, we must not lose courage in medicine nor should we despair, for the creator of enemies has also created enemies against them. There is no disease with the power to kill, for all diseases are curable without exception, even those illnesses which we do not understand.

And in *De Morte et Vita* we explain what death is. We do not intend to describe the making of these sedatives and calmants; they are perfect in themselves against such diseases. Therefore they do not need any improvement. The sedatives must have some preparation, but no other than that which we shall give in the prescription about them. Those are our first prescriptions, and they need a good artist who must not only be well read but who must be even more skillful, as we have said in our *Archidoxis and Quintessentia.*

Much depends on what is shown and taught by experience; therefore, we cannot describe it in such detail here. This is something that has not been invented by the physicians, but by artists who are the initiators and conquerors of all subtle remedies. They are not called remedies or medicines but *arcana,* because of their wonderful and noble virtues. In our opinion it has not yet been shown what these qualities and virtues are, but we shall quote this *arcanum* for the cure of falling-sickness and shall say nothing of its other virtues. It should be kept in mind that preparations of it are manifold and varied, and its virtues are as manifold as its preparation. Not each *arcanum* has this virtue; the greatest power, however, lies in the preparation and depends on the detail of the work. It is the same with the *simplicia,* if one must be more effective than the other. One is *vitriolum,* which comes from Hungary, Cyprus, or Rome, which has been especially well graded and purified. Do not feel displeased or indignant at the force of the vitriol, for it has a certain secret quality, which is not physical but spiritual. It contains more excellent virtues than gold, as we have said in *De Generibus Salium Terrae.* The lauded *spiritus* of the vitriol is subtilized and separated from the dross in such a way that there is only a quintessence of vitriol and *arcanum,* which is comparable to *aurum potabile.* It is made in the following way: take well-graded, good vitriol (as we have indicated), as much as you like, but at least five pounds; put it into an earthenware retort which should be fireproof and so large that one third of it remains unfilled. This retort should be put on a very well-made *athanar,* one well covered with clay;

put on it a well-made *alembic,* hermetically covered with clay, and in front of it put a large receptacle, preferably with a pasted beak. Then, when it has dried, make a fire, slowly as is the alchemist's habit, in order to notice the signs in the spirits and drops and to bring out all watery fluid. Drive the *spiritus* out with a mighty fire which usually takes two days and nights without interruption. And even though much studying and instruction is necessary, we do believe that enough has been said to the intelligent physician. Those who are not clever enough had better not understand it at all. Note, that all virtues go into the athanor from the vitriol, and that many mutations, spirits, and colors appear, due to the vitriol's nature.

Sometimes copper in the retort melts into a log; this is solely due to the vitriol. Sometimes it is the same with much gold; sometimes the entire vitriol melts into gold, sometimes it burns into red or black coal. Now, when the distillation has been completed, take a glass phial, pour in this distilled vitriol water, crush the *caput mortuum* into small pieces, put into it the distilled vitriol water; distill it again, so that all the glass melts together. Thus it inflames in that distillation the right and true effectiveness of its essences and the power or virtue of penetration. Then take it and separate it in *aqua forti* and the phlegm will disappear; an oil will remain, which is called *oleum arcanum vitrioli.*

Note that we have great confidence in the phlegm which has been separated in this way, almost as much as in the *oleum,* for it emerges endowed with many virtues. We have also been accustomed to apply a mixture of both of them. Both have come to us from the *arcana.* The dose is half a scruple in good wine or water, every morning and night, until the convulsions subside. The oil is sufficient to cure falling-sickness, and we shall leave it at that; for it would be untoward to put here what we have learned about it, what has happened to us. There are still other ways and manipulations for the making of *arcanum vitrioli*: for instance, let the retort stand on a fire for several weeks; or use lying flasks, retorts, putrefaction or much distillation; or use much *aquae ardentis,* imbibitions, calcination, and similar things which we all praise and like. We do not deem it necessary to write about it here, for every good and honest physician knows enough about this and similar things.

Now we shall have to leave the methods and shall proceed to the prescriptions as put in the beginning of the chapter, and this is what we say: *Arcanum vitrioli* should be taken as a base into which the following ingredients should be put: *camphore, spodii, rasure cranei, unicornu, santalorum alborum, rubrorum, citrinorum ana unciam unam, corallorum, visci quercini, granorum peonie, radicum peonie ana uncias duas*. These pieces should be crushed and pulverized to the smallest possible fragments and should be put into one pound of *arcanum vitrioli;* then let the concoction rest for a month, in the above-described way. We consider that *arcanum vitrioli* is sufficient to heal all epileptics. Therefore, no prescriptions, *composita* or *simplicia* are needed. We use this compound because vitriol is of uneven quality, sometimes evaporated or not well graded; sometimes it has another blemish which causes an obstacle, so that its virtue is less than when it is perfect. Such damage may be caused by smoke and fire. Sometimes the alchemists make a mistake in its preparation, which is the root of the fault.

Now we have to understand the way sedatives should be added. They are more useful than one might think, as for instance *opium thebaicum* and others. The preparation and mixing of the others is done as follows:

Take *opii thebaici* 2 drachms, *cinamoni* ½ ounce, *musci ambre* ½ scruple of each, *corallorum* ½ ounce, *mandragore* ½ drachm, *succi hyosciami* 1 drachm, *masticis* 3 drachms. Mix these, crush them, prepare a lozenge and add stewed hop-juice. Put the compound into a quince. Close it again, then put it into a dough and let it bake in an oven as if it were bread. Take it out and crush it, one half-ounce to five ounces of *arcanus vitriolus*. Now we have completed the whole cure for falling-sickness, and although there are still many more *arcana, magisteria, elixeria,* etc., such as *aurum potabile, aqua vitae,* etc., which are all useful in the treatment of falling-sickness, they will not be described here, as we have mentioned them in *De Quintessentia,* and we shall leave it at that.

De Cura Maniae

THE SECOND CHAPTER

We must speak about mania in the same way as we have spoken about falling-sickness. There are two remedies which remove mania, one surgical and the other physical. It often happens that both are used. In the first place, we want to mention the surgical practice, and afterwards the physical one, and then also report on the experiments which belong to each. We think surgery difficult and truly advise that nobody use it unless he be well instructed in surgery and informed by his own experience in all incidents, which we describe in Libro *De Externis Curis*. This is our practice of surgery: first, make an aperture at the place from which the mania rises; but if you are in doubt, open all extremities—at the toes, fingers and the head—in a circle which must be as wide as possible. This aperture is twofold: one which only lifts up the skin and removes it, so that the raw flesh remains underneath; the other which makes holes so that a scurf falls out and a hole remains. Now the use of one or the other should depend on the disease and thus, if the disease is strong and intense, the first opening of the skin is not sufficient to give an outlet to the origin of mania, but the second opening can, because it leaves more holes; each separate hole has the capacity to serve as an outlet. Therefore, both methods of aperture are given herewith:

First Aperture:

Take *radicum flammulae recentium et aceto imbibitarum florum flammulae recentium in aceto imbibitorum ana lotos duos, pinguedinis Meilaender keferli unum lot, cantharidarum drachmas sex, fermenti, aceti ana quantum sufficit.* Crush them all and mix them well into an ointment. Spread it on a cloth over an area as large as your fingers can stretch and if it begins to dry, wet the cloth from the outside with vinegar, so that there be moisture; let it remain there for five to six hours, then take it off, and cut the blisters open and take the skin off, and you have the raw flesh on that spot. Now we shall describe the other aperture to be used, if the first cannot be manipulated. It follows herewith:

Take *two half-ounces of aqua fort, salis communis soluti ½ ounce, mercuri sublimati 3 drachms.* Mix them together in a glass, put the compound on a small fire, let it dissolve in water and wash the width of the extremities with it from one to six times. Then let it dry and the skin will come off, like powder. Each of the described apertures is sufficient for taking off the skin. We have sometimes taken off the skin and peeled it from the flesh; this we liked best, because the blood comes out with it which soothes mania to a great extent. Some have also had their extremities pricked open with spring-lancets. Pricking, however, does not seem to be sufficient to take out mania. Only a total aperture will do this.

If you do not like to apply the described apertures, you should use the ulcerating aperture, which is done in the following way:

Take *mercurii sublimati, arsenici puri, aquae fortis, ana.* Mix them into an ointment and tie it on the extremities until the sensitiveness subsides. Then mollify it with grease, so that the scurf comes off. After it has been removed, the other practice follows. If you have made these apertures, remember that the flow of mania should be extracted with all its substance, as we have explained in the chapter on *Mania.* This happens only by means of an *attractivum,* which must be put over the ulcerated extremities twice a day. It absorbs the illness, and in the end, after this absorption, it heals together as follows, and this is how the *attractivum* should be made:

Take *galbani, oppopanaci, serapini, bdellii, ammoniaci, ana* 2½ ounces, dissolve the mixture in *aceto* and stew it in its juice; then mix it with the following well-ground, finely strained species: ½ ounce *masticis,* 2 drachms *thuris,* 1 ounce *carabe,* half a handful of *ungule caballine,* 3 drachms *magnetis,* 2 drachms *colophonie,* 5 drachms *firnissi, fiat cerotum.*

Put that ointment on and fill the holes of the extremities well with it; renew the bandage, until you can perceive a decrease of the mania. Then let it heal together in the usual ways of surgery as we have said about apertures and *attractiva*. Note this remarkable thing: that the apertures of the head should be made last, after all the other exulcerations have been closed. Only if the need arises, open the head, too, and proceed in the way we have described. Thus we have instructions for curing mania in the indicated way by surgery. It should be remembered that there are also some veins which should be opened and bled. It happens that by such bleeding mania disappears, but we do not want to say more about it, as it is commonly known.

Now we shall continue and speak about the treatment of mania in the physical way, without applying surgery. Two phases of this cure should be remembered; one that cools off and congeals the noxious matter of mania, the other which soothes and kills the matter from which mania is born. At first we shall speak of those which cool and congeal mania. This may be done externally by application or internally by medicine. We describe the greater part of it in the chapter on the treatment of falling-sickness, and this is the description:

Take 1 ounce *olei camphore*, 1 drachm *olei musci, commisce*; give the patient half a drachm to be swallowed at once. It removes mania in a miraculous way, extinguishing the heat of the body and the raging of the blood and coagulating the matter, like water which freezes into ice. We have never come across a similar prescription. Such oil can be applied for external use too, on the temples and forehead. There are still many other remedies which take mania away by their great qualities and comfort, such as *quinta essentia argenti, quinta essentia saturni, solutio cristalli, quintum esse martis, quinta essentia antimonii, solutio corallorum,* and other appropriate things such as *extractio camphore, extractio solis,* etc. There is another way of curing mania by remedies that deaden it, so that it cannot be felt any longer. There are many of these and they may be used both internally and externally. Those which are made in the same way as quintessence are the best: for instance, *summum anodinum, quintum esse mandragore, quintum esse opii, quintum esse papaveris, essentia lollii, hyosciami.* These are strong and powerful medicines; they are purified and cleaned out by the qualities of the quintessence to such a degree that no blemish or impurity can be found in them which might cause damage to the body. They raise and extract the substance gently. This is not the place to reveal their virtues or to tell how they remove such grave and incurable diseases, but this is sufficiently disclosed in *De Quinta Essentia.*

We shall finish about the cure of mania here. By this and by the medicines much will be learned, and much has been said about its cure. By experiments with herbs much has been discovered and invented against mania. We shall not burden ourselves with those, but apply quintessence and fluids which have always proved very useful.

FROM THE FIFTH CHAPTER OF THE SECOND BOOK

On the Treatment of Vesanias

Now we shall write the fifth chapter, about insane persons and their cure. There are four kinds and we shall also give four cures.

First we shall speak of the *Lunatici;* in this cure we prevent the attraction of the moon by *confortativa,* so that the *Lunatici* may offer resistance, like a roof which is put up against the sun, lest what is lying under the sun be disturbed in its existence. First, we must bear in mind that the power which the moon, as well as that of all other planets and influences which take strength from our bodies, holds over us, as well as the sun, can be taken away by the power of medicine. The moon is like a magnet which attracts all iron and steel. The power can be taken away from it as from the iron, for iron that has been covered with *oleum mercurii* will not be attracted by any magnet, nor will a magnet rubbed in leek ever attract anything. Thus we have to distinguish several medicines to use against the moon, some against Mars, some against the sun, some against all planets. Therefore, one should try to build up resistance against these influences by medicines which could be applied in correlation with the power of the moon, or with that of the other planets and stars. We write about it in more detail in *De Quinta Essentia et Influentiis*. There are seven planets. Therefore, if a planet destroys a *corpus*, the quintessence of the metal should be applied against it. For instance: *quintessence solis* against the sun, *quintessence lunae* against the moon, etc. It must be understood that *quintessence solis* is powerful against all planets, because of its *specifica* and the great power which it gives to the heart, by which all this is driven out, as we say in *De Septem Membris* and as has been reported sufficiently in *De Lunaticis*.

We also want to describe the treatment for those persons who have brought insanity from the mother's womb as a heritage. There are two cures: one is a preventive for the father and mother, so that insanity may not affect the child; it also is for the insane person himself. This first cure should be performed as follows, and be called rather a preventive or expulsive remedy than a cure: The parents should not perform a natural coitus but an artificial one. When they have the desire for intercourse this insanity is hidden and appears due to coitus, if the coitus has been performed during insanity. It leads to insanity. Then the child will be insane. If, however, no coitus is performed during the illness, but if coitus preceeds the illness, the child will not be insane. From this it follows that one should not cohabit when feeling the desire for it, but should immediately jump into cold water; thus the desire for coitus will be expelled or extinguished. When the fervor has been extinguished the coitus should be performed artificially. It can be induced and stimulated by medicine. Then a natural act will follow consistent with nature and not with the spirit or the will of insanity. Thus, whenever one has the desire for coitus, it should be incited by medicine. It can be seen by this that nature in itself does not produce insanity, but is good. If the insanity is of a permanent nature, then the coitus can be prevented every day in the above-described manner, and if the first child is not perfectly free from insanity,

his children will be free from it through this method. It should
be remembered that insane persons must be protected by quintessence before the coitus. In this way the genital organs are
protected against things unsuitable and inconvenient, so that
no evil birth or insanity may result.

The second cure is for the insane and is done as follows:
through *confortativa* or through sedatives. The cure is not
possible and cannot expel insanity unless the make-up and
humors of these persons are changed and redirected, so that
the new make-up may be stronger and more powerful than the
old. Then nature feels such assistance that all these die away.
Confortativa should only be made of quintessence, like *quintessence solis, perlarum, argenti, corallorum, antimonii, vitrioli,
sophie,* etc., and the sedatives should be made of the quintessence of sedatives like *mitigatium magnum, anodinum tem-
peratum,* etc.

By these indicated remedies insane persons can and should
be completely restored, so that they will not be insane again.
There is no other way of fighting the cause of insanity,
although there are still many other things which may be used
against it, which we shall not name or enumerate.

In the third place we shall treat *Vesani,* who by food and
drink lose their reason and become disturbed, as we say in the
chapter on the *Deprivation of the Senses.* These should be
cured in the following two ways: one in a specific way, the
other by sedatives. By " specific " we understand that the
medicine should be of the same substance as the food. The
medicine should kill the poison of insanity before it has been
taken. This can be understood in the following way: if a
person is afflicted with an illness from cat's brain, the medicine
should be such that it kills the cat's brain; there are certain
herbs which kill cats as soon as they are smelled or eaten by
them, as they have a specific effect on their brains. Thus, if
a person is taken ill by cat's brain, he should be cured by such
herbs. If someone falls in love through a potion, that love
should be destroyed in the same way, so that the potion be
driven out. We shall not continue about specifics, for we do

not wish to disclose how people can be contaminated, and therefore we stop writing about remedies against it. The sedative cure is achieved by quintessence, which alone has the power to remove and soothe. The sedatives are: *aurum potabile, quintum esse lunae, perle, opiata,* etc. It must be taken into consideration that, as we have said in *De Choreis,* it is not possible by ordinary prescriptions to find in nature a perfect medicine for each individual. Therefore only quintessence takes away and expels such *vesanitas,* and it should be mixed with quintessence *opii, mandragorae, hyosciami, lollii,* etc. The quintessence should be of gold and silver, chelidonia, etc., as is especially indicated in *Quintessentia* and in *De Gradibus Maioribus Super Quintessence.* Because there is no remedy which can remove food or bewitched potions, one should not purge the body nor remove the poison in such a way, for it is of no avail, as the insanity lies in the *spiritus vitae,* which cannot be affected by any purge. Therefore one should use only *confortativa,* so that the *spiritus vitae* should not be overpowered and settle or die.

The fourth group of insane persons is that of the *melancholici* who are disturbed by their own nature—there is no apparent defect of reason; their complexes are affected and they suppress reason, ruling it as they wish. There are two questions to be considered in the cure of melancholy. First, from what complex it originated, and second, how it can be expelled. They can be understood as follows: if it is melancholy, then one should apply *contraria.* If the melancholic patient is despondent, make him well again by a gay medicine. If he laughs too much, make him well by a sad medicine. There are some medicines which make people laugh and make their minds happy, removing all diseases which have their origin in sadness. This is not incidental, not just laughter in sadness: the entire sadness is removed. There are also medicines which induce sadness, in such a way that they soothe unseemly laughter and exaggerated, unsuitable pleasure by changing it. Thus, the reason is set free and the memory is completely normal. In this treatment it must be remembered that this medicine can be

made from quintessence alone, which has such tempering qualities that it leads nature back to the right course. The following
medicines should be known, as they serve against melancholic
disease and expel all sadness, or free reason from sadness:
*aurum potabile, croci magisterium, arbor maris, ambra acuata,
letitia veneris.* They cause insane patients who show excessive
pleasure and voluptuousness to be naturally sad. Although there
are four kinds of melancholy, originating from the four complexions, there is no need to mention specific medicines for
each separate complexion, as the two complexions mentioned
will suffice, for two complexions are alike: *sanguis* and *cholera*
are accompanied by joy, although they are different; for one
is warlike, the other is not. They have the same cure and it is
sufficient for both. The other two complexions, phlegm and
melancholy, behave in the same way; therefore both can be
cured in the same way. And now enough has been said about
these melancholic patients.

On the Treatment of St. Vitus' Dance, or Chorea Lasciva, or Recklessness of Mind

We shall describe what may seem an unusual treatment of
St. Vitus' dance, according to its causes. *Vitista* inherited its
name from St. Vitus, who is supposed to send such a disease
to people as an ordeal; it may also be sent by God, because of
people's sins. We do not know whether there is some truth in
this, but one should not think that this disease is a plague
and that God or the Saint inflicts it. Rather, they impose it,
or permit that by swearing St. Vitus' name such an imagination
may appear, bringing the disease with it. The common people
consider this a plague, sent by the Saint, but it really is nothing
but an imaginative sickness, as we have said in the corresponding chapter, and we have a special cure for such imaginative
vitista. Accordingly, if it originates from careless spirits and

impaired will power, then it is called lascivious disease (*chorea lasciva*) or recklessness of mind (*levitas animi*), and we prescribe another special cure for it. Against the third type, *chorea coeca,* which has its origin in nature, we prescribe its natural medicine. This practice, therefore, should be understood in three ways, since no other cure would help against it, and each requires its own cure.

The first is the *cura choreae estuationis,* or *imaginationis,* which comes from swearing. There is a difference between the type of sickness which comes from rage and that which is due to voluptuousness. Therefore, we consider the first, which comes from rage and swearing, and divide the treatment into parts. It is as follows: the patient should make a likeness of himself in wax or resin, and should concentrate on it so that all the curses he has uttered may be destroyed in that likeness, by his will. He should concentrate his mind and the memory of his swearing solely and entirely on that likeness, thinking of no other person, and then he should cast it into the fire, letting it burn completely so that neither ashes nor smoke shall remain. In this way the thoughts pour so strongly and powerfully from him into the likeness that they cannot be directed against him, as if the image were alive. For curses work against those who utter them and not against the men at whom they are aimed. There is no resistance in the image, but it is physically destroyed and the thoughts are destroyed with it; so that this may be well understood, we give the reasons in *De Imaginationibus* and in *Corpus Impressionibus.*

Now we have said enough of the *cura vitiste ex ira,* and we shall begin with *chorea lasciva.* We know from experience that one force expels another. Therefore, we shall give our experience here, so that *lasciva* and *intemperata* can be expelled. It is done as follows: if a choreic man or woman develops St. Vitus' dance through a voluptuous urge to dance—and this happens more frequently to women than to men, since women have more imagination and restlessness and are more easily conquered by the very strength of their nature—there is nothing better for expelling this than thoughts and actions against it. Their thoughts are free, lewd and impertinent, full of lasciviousness and without fear or respect. These thoughts may be expelled in the following way: shut the patients into a dark, unpleasant place and let them fast on water and bread for some time, without mercy. Thus hunger will compel them to adopt a different nature and different thoughts, so that the lasciviousness is driven out by abstinence. This is the best way, since their immodest actions and blood settle down, the *spiritus vitae* changes and slows down, and the heart becomes sad because of the sadness of the place and the changed, imprisoned way of living. So the old disease vanishes and sadness becomes the master. There is no joy, no laughing, no dancing, no howling. When this result has been accomplished, then you should begin to mollify the hard life a little day by day, to improve the food a little every day, to ease the confinement until melancholia too has been consumed, so that the patient may come back to his senses. This prescription is against the thoughts and actions of these people, and it is foolish to give in to such dancing and to their will and way of life, with its singing and dancing. Such giving in only stimulates and furthers the disease. Some think they would die if they could not act in such a way (singing, dancing, etc.) but it is not so. It is better to take a good stick and give the patients a good beating and lock them in as has been described above. It should be noted, however, that if they are beaten, such a rage arises within them that they may die of it; therefore, one should be careful to observe moderation. The best cure, and one which rarely fails, is to throw such persons into cold water.

Now we must speak about the cure for *chorea naturalis,* which comes from nature; we shall show how it takes root in the laughing veins and accordingly causes dancing and jumping. We shall describe the treatment in two ways, according to its origin: an internal way and an external way. By external way we understand proper localization of the disease. The internal medicines from quintessence are these: *aurum potabile, aqua margaritarum, quintum mandragore, aqua vitae, anodinum summum, quintum papaveris, materia gemmarum, quintum opii, quintum esse lollii, cancrorum,* etc. Some of these medicines should be applied not only externally but internally, as for instance *summum anodinum, essentia mandragore, essentia opii, quintum esse lollii.* An ointment should be rubbed into these parts. The following has been our practice: take one drachm of *essentia opii,* seven grains of *essentia mandragore,* one scruple of *essentia lollii,* two drachms of *papaver,* three drachms of *hyosciamus,* one half-ounce of *sol potabilis,* three half-ounces of *aqua corallorum.* Four drops of this mixture should be dispensed in good wine or in water until the end of the cure. An ointment made of one half-ounce *summi anodini,* one ounce of *oleum muscate,* and one drachm *olei musci* should be rubbed into the extremities and particularly into the ticklish places, like the armpits, throat, etc. The same treatment should be followed for those who are locked up, but not so strictly as advised in *De Lascivis,* so that some sadness may be introduced and a change of the body brought about. We shall end this chapter, as enough has been said on the cure of *chorea.*

FROM THE FOURTH CHAPTER OF THE SECOND BOOK

On the Suffocation of the Intellect

In the fourth chapter of the first book we spoke about the origin of *suffocatio intellectus.* In this fourth chapter we shall describe the treatment and cure of *suffocatio intellectus et sensuum.* We mentioned three kinds of suffocation and shall now describe and examine the first of them. It is understood that suffocation originates from worms.

Take an ounce each of *colocinthidis, esule, harmelli* and *hellebori nigri,* three ounces *agarici,* one half-ounce *scammoneae,* six ounces each *polipodii* and *sene, mellis despumati, vini optimi,* three pounds of each, *aciti modici adde.* Boil all these ingredients together and when they have been stewed long enough, strain them. Give a large drink of this—depending on the severity of the illness—to the patient. When the use of the laxative has proven effective, give the patient another drink after a short interval and thereafter as often as you deem it necessary, for this medicine drives worms and their nests away by its quality and by its *specifica*, and no worms can grow in it. Also, if you apply those medicines described in *Contra Caducum* in the same dose, it takes away all paroxysms in old and young persons, and you will never find a loss of reason due to worms. There are many other *appropriata*, *specifica*, and *empirica*, which belong to this group and serve this purpose, but we omit them since we do not know of them from our own experience. We trust those who describe these others, and we mention only those which have come to us through experience and which have been tried out by us.

Now we shall describe the cure of the suffocation which afflicts women from the womb, as has been mentioned in the fourth chapter of the first book. The cure is performed in three ways: one by externally applied ointments, the second by fumigating, and the third by intake. An ointment can soothe the disease by its heating power, but it does not completely remove the disease. It is an excellent help in the two other methods of cure and is prepared in the following way:

Take *olei tapsi ex solis floribus tapsi, olei anethini ex solo viridi anetho ana unciam, olei olivarum et amigdalarum ana unciam dimidiam, balsami, masticis, drachmas duas.* Mix the ingredients together and rub the resulting ointment in around the navel and the pit of the stomach; also, put on the belly a bag containing previously well-stewed *artemisie, camomille,* beech tree ashes, five handfuls each. If after prescribing this medicine you see that the suffocation will not subside, but is almost like falling-sickness and looks terrible, then conduct fumes from this medicine through a tube to the womb. This alone will be sufficient to remove all suffocation of the womb, even if the patients have been near death.

Take *verrucarum equinorum genuum unciam unam, assae fetidae drachmam unam, cornuum et ungularum capre drachmas duas.* Crush them and mix them into a powder, put one scruple over a fire and apply it in the usual way. This draws suffocation out of the heart, clears the head and brain, cleans

the womb and purifies the body more than you can say. Note
that if you do not notice and feel sufficient relief, an internal
medicine should be used. For it happens quite often that nature
is destroyed by the convulsion so that it cannot resist the noxious matter without *confortativa*. And even if the suffocation
becomes less or subsides, you should still apply those *confortativa*. If you cannot obtain them, there are the following
confortativa matricis which suffice to strengthen nature:

Confortantia:

Take a quint each of *aurum potabile* and *materia perlarum*,
half a drachm each of *quinta essentia anodini* and *quinta essentia
hyosciami, misce simul;* dispense one scruple, to be drunk at
intervals of six hours.

Confortantia Matricis:

*Materia perlarum, materia corallorum, materia cristalli,
quinta essentia opii, quinta essentia papaveris, quinta essentia
hyosciami, et sic de aliis similibus contra colicam.*

It should be borne in mind that if the suffocations are not
permanent but afflict the patient only at certain times, they
should be prevented in the manner described in *De Preserva-
tione,* which is omitted here.

We shall also describe the cure of the third kind of suffocation, which is as severe as all the others. It has several aspects
and by this cure we shall cover all kinds that have been described in the chapter on suffocations, for they are cured by a
similar treatment. No special medicines are necessary for it,
but only those which have been prepared from quintessence;
this is so for many reasons, but above all because the diseases
come from the weakness of human nature; therefore they need
the medicine which strengthens human nature. By the strength
of it all suffocations are destroyed. We shall therefore indicate
those which act against suffocations and have been mentioned
under *quintessentia.* They are:

Fortiora:

Aurum potabile, appropriatum sanctum, succus perlarum, maximum simplex, arcanum vitrioli.

Mediocria:

Salis oleum, aqua aurifera, materia perlarum, oleum lune, aqua specifica, materia corallorum.

Minora:

Oleum martis, oleum de venere, oleum saturni, oleum mercurii, oleum iovis, oleum christalli, oleum sanctum, quinta essentia sanguinis, quintum esse coriandri, quintum esse peonie.

Grandia Repugnantia:

Quinta essentia opii, quinta essentia mandragore, quinta essentia lollii, quinta essentia papaveris, quintum esse gemmarum, quintum esse tartari, quintum corallorum, quintum sulphuris, quinta essentia anthimonii.

Now we shall conclude the methods of cure. The treatment cannot be improved and is not to be changed, for by its virtue not only suffocations but even death can be expelled, as we say in *De Quintessentia* and *De Mente et Vita*. We shall leave it at that and now proceed to other things.

FROM THE SIXTH CHAPTER OF THE SECOND BOOK

On Prevention

Now that we have finished with five chapters, we shall speak of prevention. It would be a good thing if there were a special remedy for each individual kind of insanity. This, however, would take too much time. We shall provide ourselves with a prevention that protects man against the first loss of senses in epilepsy, and also against *mania, chorea, suffocatio* and *privationes sensuum*. Such a remedy cannot be produced without great skill in separating, for the remedy can and has to be brought about solely by quintessences, which by their miraculous strength prevent all these deprivations of reason. Not a little virtue and strength are needed to beware of and to prevent such grave and overpowering diseases. Therefore, we shall mention those quintessences which master *privationes sensuum*. Although such diseases can also be prevented by minor medicines, such as *simplices* and ordinary compounds, and by good order and regularity, we shall not describe these remedies here, as they do not often help. This is not the case with the quintessences, unless the disease is a *morbus hereditarius,* as we say in *De Sanitate et Aegritudine.* There is no remedy against this.

The quintessences are: *sol potabile, oleum solis, oleum lune, oleum ferri, oleum mercurii, oleum saturni, oleum iovis, oleum veneris, oleum vitrioli, oleum tartari, oleum anthimonii, quinta essentia auri, quinta essentia argenti, quinta essentia ferri, quinta essentia mercurii, quinta essentia veneris, essentia saturni, essentia iovis, soluti coralli, soluti christalli, solute gemme, solute perle, essentia melisse, chelidonie, carline, florum utriusque helleboris, opium magnum, anodinum summum, essentia mandragore, essentia opii, essentia papaveris, essentia lollii, essentia hyosciami, aqua vitae aurea, caponis sanguinis, elixir tincture, tinctura,* etc.

The reason for the strength of these medicines will not be given here, but will be stated in the book *De Quinta Essentia.*

Now we shall praise God Almighty forever and ever, come to a close in the name of the Holy Trinity and end the book *De Amentibus* for the benefit of the common believers and nonbelievers, the evil and good, the poor and rich, so that wrong may become right and right remain right. Amen.

FROM " LIBER NATURALIUM "

On Vitriol

Nature brings forth a salt called vitriol. This vitriol is a special salt, different from all others, and it has more virtues and qualities than other salts and such high and great virtue that it is appropriate to mention it in this book. For vitriol offers a complete cure of jaundice, sands and stones, fevers, worms and falling-sickness. It is a great deoppilative against constipation of the body, and has other virtues too which will be discussed in the other chapters. It will be described with regard to the two arts, medicine and alchemy. In medicine it is a miraculous remedy and in alchemy it is excellent and useful for many other things. The art of medicine and alchemy lies in the preparation of vitriol, for when it is crude it is not in a suitable condition for use but is like a piece of wood from which you make many pictures and carvings. Vitriol is excellent as a medicine for the body; it can also be used for external application, for wounds, and for surgical diseases such as hereditary scabies, leprosy, ringworm, etc., against which we have no other remedies and which cause death. These are vigorously attacked by vitriol, which cures them from the root.

Therefore, the physician should know how to prepare vitriol, as it has several virtues in its crude condition, some when it is turned into water or chalk, some in the form of green oil. Its quality is that it reveals new and special secrets whenever it is transformed into another form or essence. It puts all Italian and German apothecaries and all their writings to shame, for it is a medicine which in itself is sufficient to fill one fourth of all dispensaries and is satisfactory against one fourth of all diseases. Therefore, those ordinarily numerous boxes, containers, cases, pitchers, jugs, and glasses are superfluous in the dispensaries, as they contain worthless medicines and can do little good and even some harm. The physician should try not to depend upon the number of boxes and medicines from faraway countries and not to look so far from home. He should cast his eyes down like a maiden and then he will find at his feet more treasures against all diseases than we could find in India, Egypt, Greece, and other foreign countries. For there is only deceit in the boxes and cases, and their doctors and apothecaries are as wooden as they. Like attracts like.

The Kinds of Vitriol

There are many kinds of vitriol. In each pit there is a different kind. The test of whether it turns iron into copper is not sufficient. The correct test is whether it can be tried against worms, as we shall say later. It is excellent in medicine, because it drives worms out. In alchemy there is this test and not only that of turning iron into copper. But the latter is a proof too. Vitriol makes copper out of iron. If it makes a great deal quickly and of such a high grade that it is soft under the hammer, the better it is for medicine and alchemy, for it means there is such a union between iron and the vitriol that the vitriol turns iron into good, lasting copper; the best copper comes from good vitriol. Nobody should be amazed at hearing that copper is made of iron by means of vitriol, for another of nature's powers is that of making living mercury out of lead by borax water. Also "Kakymies" turn some metals into others, just as the vitriol does with iron; for vitriol is not the only transmuter of metals, but there are several transmuters for each metal. It is the fault of the sophists that we do not know all about it and that these arts have gone into the kitchen and the money bags. In Hungary there is a brook full of vitriol and it is a vitriol that is not coagulated in pebbles. It will eat up all iron you put into it, turning it into rust. This rust is put into a furnace and then it becomes pure copper, which is true copper and remains copper and does not disappear. There are countless vitriol ores in Germany, so many that you cannot separately describe each kind. In alchemy and medicine, however, the test has to be made in the above-described way and one should act accordingly. There is another proof not helpful in medicine but only in alchemy; that the *colcotar* of vitriol makes copper out of it by means of fire. Also, the colors of vitriol should be noted. Those that are marked with blue and no other colors are not so good as those surrounded by red and yellow or covered with it, and that which turns white in the air is good for the white and green oil of vitriol; the one that turns red or yellow is better for red oil.

Therefore, we need not say more about the species than that the alchemist should act in conformity with the test and everything it shows if he wants to use it. There is another test which is also good: if the vitriol becomes almost black with *gallales* and yields dark black ink, it should be preferred to all others. If it gives watery ink and needs many compounds, it is not very good.

On the Virtues of Vitriol in Medicine. On Crude Vitriol and Colcotar

Continuing to report on vitriol and its virtues, I shall have to show you above all the virtues of true vitriol and then those of its colcotar. We know that vitriol is an excellent purgative for serious stomach diseases. It often happens that by food and drink the stomach becomes so upset that a long illness develops and remains until death. It occurs mostly with those who eat and drink irregularly. They fall ill and sometimes even die from such ailments. Also, in wartime there is much irregularity, etc. Sometimes the illness is followed by contractions, fever and other disease, and it is obvious that then it is best and most advisable to purge with the vitriol which has been secretly named *grilla*. The dose is six *comes*. This is enough if dispensed to a weak patient in wine or water, or, to a strong one, in brandy. It tears the disease out by its roots from head to heel. The doctor should also bear in mind that hellebore, ergot and devil's root as well as *coloquint* have a quick, strong, purgative effect, but not as much as vitriol. The excellence of vitriol is caused by its great bitterness; it contains an acidity, sharpness, and thorough cleansing power which can never be found in hellebore or *coloquint*. There are two methods of purging. The first purges by itself, the second by " salsoditet." This is a specific form, which makes all the worms it touches die; none of the remedies which have been mentioned above ever does it. This is a great virtue. Therefore, you should know and keep in mind that a purgative which contains bitterness and salts not only has a laxative effect but cleanses the body thoroughly. Therefore, it is doubly preferable to other purgatives in such diseases.

About the colcotar, however, you should know that it can be applied in surgical cases such as in purulent wounds which do not seem to heal. Colcotar is the basis for their healing. But, as you know, there are many external surgical diseases: high and low, bad, worse, and still worse. Colcotar has no power over those which are worse and still worse, but only over the first stage. It should be applied only depending on the condition of the sore.

It therefore follows that all sores can be healed by vitriol, but always depending on their degree. The vitriol is usually made into a good colcotar, which should be soaked in vinegar several times, then dried and strained or mixed into a simple ointment, depending on what the physician deems necessary. A scurf is left which should be removed with fats and then a good recovery follows. If a wound resists the healing process, then you should know that there is some surplus poison and that it must be treated with oils. That is the reason why colcotar may be considered insufficient for a complete recovery. If crude vitriol is dissolved in vinegar and made into a powder, together with the colcotar, and applied in this way, it offers a much stronger basis than that of the mere colcotar. The best way of producing a colcotar is to take the water out of it and to soak the *caput mortuum* in it until it has absorbed everything; then let it dry gently in the air. This when applied yields a better result. But always watch the degree of the sore, ringworm or *sirei*; those which do not respond to treatment must be cured with separated oils and water, as we shall describe later. You must not despair if the colcotar fails in a cure. You should try to improve its distillation, for only by preparation can the medicine be brought to its highest degree of effectiveness so that it is able to offer resistance to any surgical ailment, *lupus,* cancer, *sirei,* as we shall indicate when writing on such preparations.

On the Use of White Vitriol in Physical and Surgical Diseases

Many great things have been revealed to the physician by the art of alchemy, so that excellent cures for all diseases have been accomplished. Therefore, even in the beginning of medicine all physicians devoted themselves to alchemy, because it brought such praise and usefulness to medicine. The two arts, medicine and alchemy, have again and again emulated each other and crossed each other's path until the drivelling sophists and humoralists appeared. At that time poison was poured into medicine and medicine became the whore she will be so long as there are humoralists. I am telling you this because you should pay strict attention to this chapter because of its great value.

One instruction I shall have to give you first of all: when ignorant and uncomprehending people intrude into an art, they spoil the whole thing and make a manure heap out of a well, as has been done with vitriol. At first people seized upon the spirit of vitriol and valued it as highly as possible. When that had been done, they cured falling-sickness in old and young, in men and women, in all kinds of people; but now the ignorant chemists have appeared and want to improve it. They dared to force vitriol and its virtues into another direction; they let the *arcanum* boil over so that it evanesced; they searched for the oil in colcotar when oil has nothing to do with it. For everything that is to remove epilepsy should have a sharp, fine, penetrating spirit, and then it has the power to penetrate the body and to cleanse everything. By such searching the disease is found in the place where it is located; no man can know where it is, where and how it is located, its center, or the period at which it starts. It follows that physicians need more medicines which penetrate the whole body. Therefore the humoralists and quacks cure no one and their occupation and actions are nothing but humbug. In my opinion the oil which is sought by the chemists does not contain a penetrating spirit, but is earthly and goes no further than where it falls. Therefore wherever such foolishness has spread, stupidity has suppressed the right procedure and put the wrong one in its place. But no good has come from this, because it has not benefited the invalids, and therefore the sect of the humoralists is not successful.

Now let me tell you how the *spiritus vitrioli* was found for the first time. It happened in the following way: after having separated the humid spirit from the colcotar, men distilled, graded, and circulated it as highly as one can through this process. The water thus obtained can be used for various diseases, both externally and internally, and therefore also in falling-sickness. Patients felt signs of recovery. Therefore, men took still greater pains with the extraction, taking out the very best *spiritus vitrioli* and distilling it from colcotar in the hottest fire. The dry and humid spirits were both in it. They were extracted gradually. Then both spirits, the humid and the dry—which had been in one phial—were graded to a level. Then they gave this medicine to patients and found its effects even better than the first extract; they had such good results that all humoralists were put to shame. A correction was made by several masters by the addition of brandy in order to empower it still more, but the result was not found to be better.

I am informing you about the procedure I followed and advise every doctor to follow it, especially in falling-sickness, which is cured by vitriol. Therefore, we make better and higher efforts for the patients; these efforts are required by love for one's fellow-creatures, which a physician should have. This is my procedure: the vitriol is imbued with *spiritus vini,* then distilled as I have described above, for the dry and humid *spiritus.* I find that when this has been done, if *spiritus tartari correcti* is added in the proportion of one third to the vitriol, and also a fifth of *spiritum aquae theriacalis camphorate* to the vitriolic spirits, the result is excellent. This should be given to the patient before the attack, or several times a day. You should know that there is great power in such medicine against this disease, so great indeed that if I had one free wish, I could wish for nothing better in nature.

The first process should remain, just as it was invented by the ancients, together with the reported correction; then one can get at the heart and strength of nature. This should be taken into account: even religious people will not think that I am wrong, if they think of the great suffering caused by this miserable disease. It is lying right before their eyes, so that not only a human being but even a stone would take pity. As this disease is so much more terrible than all the other illness we can see, someone should come and say: cursed be all physicians who pass by and do not help, like the priest and the Levite in Jericho who passed by and left the wounded man lying in the road, and only the Samaritan came to his rescue. Judge for yourself what great damnation the Levite and the priest received for shutting their eyes. The physicians do the same with epileptics. Such vitriolic extractions of the *arcanum* are not only good in falling-sickness, but also in its varieties, such as in *syncope,* ecstasy, etc.; also in all constipations and internal *apostematibus* and other similar diseases and in suffocation and precipitation of the womb.

Physicians could find still other virtues in vitriol, apart from those I indicate, if they were really zealous. You should also know that the prescriptions, in which I have primarily described how to prepare the humid *spiritus vitrioli,* cannot be written more clearly; a master will understand and they will make sense to him. Also, the correction of *spiritus vini* will be understood by them. You should also look up *aqua theriacalis,* in my *Practice* in the chapter on the cure of epilepsy, and whatever else that is necessary. You should also bear in mind that all the property and strength of vitriol depend on extracting the *spiritus vitrioli* and on grading it most carefully; then, with its additions, it should be made to penetrate the body so that center, root, and germ of the disease may be reached, for there is no other way to locate them.

On the Oil of Red Vitriol

We also know how the colcotar is turned from vitriol into oil by distillation in the retort, as is known to alchemists. The vitriol turns red as blood and becomes very acid. This is now the oil, which has been sought by chemists; it is better than the *spiritus* which we have described in the diseases. The procedure is very simple; it depends on manipulations, on skill, and on good instruments. But you must know its virtues. First of all, it is more bitter than anything else. It also has corrosive qualities and, therefore, it should be applied skillfully and properly; that is, not merely by itself but in a well-dosed mixture and in harmony with the disease for which it will be used. Here is an example: *thyrus* is a poisonous snake. By itself it can do nothing, but in the composition it is excellent; *thyriax* is made of it. In the same way we shall speak here of the oil which must not be applied by itself, but only in a uniform composition, like the one mentioned with regard to *thyriax.* The acid from *thyriax* does not do any harm. It is good for the stomach in which there is no cholera or apostema; if there is cholera or apostema, no good will come from it. The apostema grows worse through it, swelling because of the acid. If there is cholera, however, then there is one force raging against the other, like tartar oil and *aqua fortis,* because neither wants to remain with the other. Therefore, good care should be taken that this be not overlooked. It is the same with the liver. In short, it indeed requires great pains.

Apart from the instances we have mentioned, its composition brings recovery in all kinds of fevers and for the irritated stomach, but each sickness requires different treatment and not too much. People say there are many virtues in the oil, but I have had little experience with it. And I have found by inquiry that those who talk about what they can do with it are actually telling lies. It is used against sand and tartar, but I have not heard of anyone who has been cured by it. It has some effect in all cases, but it is not quite perfect. In *De Cura Arenae* you will find a compound of it for use against this and other diseases. But as I have told you about the composition of the oil, I say that this composition, although it may be able to grind the stone, to crush and drive out the tartar, does it with such clumsiness that another invention will be necessary; otherwise the composition is not acceptable. But I can always acknowledge what each man knows through his own individual experience. It is a new medicine; therefore it should be tried out anew day by day; it should also be mixed with the proper compounds.

But, in connection with surgery, all should pay attention to this oil, both those in pain and in great suffering and those with good health and without disease, and they should keep in mind that this procedure takes scabies away at once, as if one were taking off an iron hat and putting it away; the oil removes scabies if, on three days in succession, you put it on with a feather; then you let it work. If, however, you see that this is not sufficient, apply more of it; according to the degree of scabies, you may also mollify the compounds by mixing it with an infusion of *herba chelidonii,* and apply it more often. But, in summing up, the patient who wishes to recover must keep in mind that this cannot be achieved without pain. For an invalid it is the same as it is with a pregnant woman who cannot give birth to her child without pain. Therefore it must be endured, unless it be changed by God. Just as we earn our bread in the sweat of our brows, we recover from illness only through pain. You should know that all eruptions have been cured in this way, and also all scabies, itches and other diseases of all kinds that gather on the skin, also *lentigines, praue, sirei,* etc. If rubbed in, the compound removes and kills what has been there. It also cures *lupus* and cancer, but great care should be taken with these. Where *sirei* has grown so that it has poisoned the sound flesh, so that by and by it will fall off and also be eaten up, *opodeldoc* should be put over it as a protective measure. Also, in other corrosive sores and where other medicines are of no avail, this one will help because of its quality. It is not a bad method of application to mix this oil with a thick ointment; it then works more slowly and mildly.

In short, you can avoid pain as little as can a woman who is giving birth to a child. It is still better if the red vitriol oil is distilled into a spirit. Then a small quantity will be sufficient. Also, in other diseases it works fast and easily, corresponding to the speed of the work. This you should know about red vitriol oil, as I have learned from experience. In an emergency it is a great and excellent medicine for healing dirty diseases like scabies, scrofula, *reüdig*, *schebig* and other such disgusting, gnawing leprous diseases. There are not many medicines acting in that way; therefore the physician should consider this oil a firm pillar of the apothecary. Concoctions of it are of no avail; they do not remove such sad diseases but only rub them in, so that they thrive and will come soon again.

On White and Green Vitriolic Oil

It should also be noted that an oil is distilled from red vitriol by extraction, sometimes white, sometimes green, which is the quality of vitriol. This oil deserves great praise. It comes from crude vitriol; therefore it contains also the crude *spiritus,* about which I have written. This oil proves in itself that, with its virtues, it is a good oil, to be praised more than any others in the treatment of the internal diseases of which I have spoken. Now let it be known that this oil, whether green or white (but the green oil is better), if it is circulated and mixed with the correction, as has been described in the *spiritus vitrioli,* will make beyond a doubt a perfect medicine against falling-sickness in all its forms. It should also be remembered that it is rare for the disease to appear alone; it has subsidiary symptoms, as we find in all diseases. It follows that the main part is not sufficient in itself; there are also the secrets of the subsidiary symptoms which may appear. Therefore, I have written in my *Practice* the chapter on falling-sickness not so much with regard to the main disease but in order to make you understand the subsidiary diseases. But if there are none, there is no reason to do anything against them; it then suffices to act in conformity with the principal disease.

Now we shall direct our attention to the green vitriolic oil. You should distill it as carefully as possible in such a way that it be separated from glands and impurities by *aqua fortis.* Then it should be taken through fire. Then the *spiritus* of the oil should also be collected and allowed to circulate in itself. Then some *spiritus vini* could be added. If the compound is to be administered solely as a medicine for the principal disease, nothing else should be added (if there is no subsidiary disease, no medicine should be added). Its dose is ten swallows in *aqua paeoniae,* always following the attack. Watch whether the attacks come in quick succession and note their intervals. As soon as the *spiritus olei* has found the center of the disease the convulsion is weakened and quiet and becomes even quieter. When the cure takes effect it makes the patient dizzy but he does not feel it. The patients do not fall, or foam, or beat, etc. They retain their reason and fall into a light sleep; the longer the sleep lasts, the more it leads to soothing recovery. At last the dizziness also subsides, yet one should not stop, but continue with the application of the medicine, depending on the disease of the person. Continue the treatment until the recovery is reached. This is the way you should perform the treatment, and do everything with great zeal, etc.

On Vitriol. On Vitriolic Oil to be Used in Alchemy. On Crude Oil

In this chapter I shall first tell you about the crude vitriol, so that I may teach you what tricks there are in vitriol. All crude vitriol makes copper out of iron. This is a natural virtue, not effected by the alchemist but by the vitriol through the operations of the alchemist. It seems amazing, in a natural light, that a metal should have the quality to disappear and to be turned into another. It is as strange as turning a man into a woman. But in these things God has given special freedom to nature, though not to man. Therefore I declare that on this point of transmutation the serene philosopher Aristotle was not very sound in his philosophy but rather possessed of foolishness. Now I will give you the prescription so that in all parts of the German nation you may be able to make copper from iron, that is, to turn iron into copper. In putting this into effect we have to realize that there may be some other transmutations which, however, are not known to us. It is not as difficult to turn iron into copper as to make gold out of iron, because God has not revealed the latter. Most things are still hidden from us till times to come, when the Saviour will come (for there is a saviour for the arts just as there is one in other respects).

But this transmutation is as follows: take one pound of iron chips without any other metal, like copper or tin, add one pound of mercury, put it into an iron pan, pour into it one pint of vinegar, ¼ of vitriol, three ounces of *sal ammoniac*. Let it boil altogether; stir it well with a wooden stick; if the vinegar evaporates, add some more, as well as new vitriol. In this concoction iron boils into copper, and when it has burned into copper put the whole into mercury; if you have let it boil for ten to twelve hours, then separate the mercury from the iron which is still there, and wash it until it becomes clean and pure; put mercury into a small fustian or leather bag, press it through and you will find an amalgam; wait till it loses its smell and you will find good and fine and pure copper. If you take one ounce of copper and let it flow into one ounce of silver, it will instantly turn into a silver of sixteen grades, and that will be the proof that this copper is made of iron. But the grades are not fixed; those who know how to work in alchemy will be able to make a living out of that; however, it depends on the art and on the manipulation, which are unknown to many persons. Thus you can always make copper out of iron by the described process; I am saying that because the transmutation from one into the other is possible. It is also of the nature of vitriol to become copper after it has been calcinized in its colcotar, with inconsiderable and light liquefaction. For there is a queer, copper-like nature in it, and also vitriol will cause a strange quality in copper; and if copper is broken by *aqua fortis* and granulated, the entire copper turns into vitriol, and can never turn into copper again. Vitriol does not become vitriol again after having been turned into copper, unless it be turned into it by means of *aqua fortis*. There is a remarkable resemblance between copper and vitriol, because they have such a way of reacting to each other. Whatever comes from copper gives a good vitriol; as an example, *verdigris* gives beautiful, high-grade vitriol in the spagyric way, and it is certain (although one should not mention it, and there is much mockery about it) that there is an excellent tincture hidden in vitriol, which can do more than many people know. Hail to him who understands it.

Now mark my words on the oil from vitriol. If oil *argenti vivi* and that vitriol are put together and if they coagulate according to their process, it gives a sapphire of a strange kind, not the sapphire stone, but one with a wonderful tincture and strange ways, of which one cannot talk. Therefore, I say that there is great secrecy in nature, as well as in other things, as in God's creatures; and at this very hour it would be better if one studied such things rather than running after evil things, etc. But this is a time which highly appreciates whoring, and that practice will go on until one third of the world is destroyed and one third dies by cheating. The last third will hardly survive. Afterwards, it will be put right again, but at present it cannot be. There will be a golden world. Man will come to his senses and live in a human way and not like an animal, not in taverns.

Although I have written so much about vitriol, it is necessary to see the miserable invalids in their falling-sickness, that every physician in his own conscience may think of God our Creator, that he may love his neighbors, and that he may not despise, refuse, or reject God's gifts which are in vitriol. For love's sake he would work at them day and night, so that nobody could be found who was lazy, etc. Then you will be so highly endowed by God, that you will be lacking in nothing toward invalids, and you will be granted everything.

Process and Ways of Vitriolic Oil by which Four Diseases Can Be Cured: Epilepsy, Hydropsy, Pustulae and Gout

Prepared in such a way as to Prevent the Errors committed by Philosophers, Artists, and Physicians.

The *spiritus* is drawn out of the mineral *vitrioli* by the colcotar, which is false and really nothing, unless it be called phlegm, which is the noblest spirit, and to which every power is attributed. Although the oil from colcotar is of great effectiveness in *harena* and *lithiasis* and in *alopecia* and other diseases, it is of no avail in the four diseases in the above title. Hippocrates, with whom all the others agree, indicates the number of years, duration, symptoms, etc. and says the epilepsy is incurable for the same number of years as gout. But as these had no experience with the *spiritus vitrioli,* this tract may be committed to the vultures. First, the *spiritus* should be extracted from vitriol over a big fire in a stirred *cucurbith* until it has reached the ninth alembic and on the hottest fire it should be driven through the reverberator in the athanor day and night until the fourth day. In this way, the *spiritus vitrioli* is prepared.

Then, on the third day, the colcotar must be distilled in a horizontally-placed phial, in the athanor, on the hottest wood and coal fire until one pound of colcotar yields one and a half pounds, colored like scarlet in the receptor. Then *alcali* should be drawn from the retort, dissolved a fourth and a fifth time, and finally it should coagulate. Thus the three things contained in vitriol are extracted and separated.

Procedure:

With epilepsy the following procedure should be taken: after each convulsion one scruple should be administered to the patient in a good wine. The following dose of oil from the colcotar should be given morning and night: four grains in *aqua de paeonia*. This procedure should be followed to the fifth attack. In cases where the attacks do not come in quick succession, half the dose should be given for thirty days.

In cases of gout the above-mentioned dose should be taken for thiry days; but continue to rub the *spiritus vitrioli* into the place where the pain is located until the gout has disappeared. If, however, it is an old gout, one quarter of *liquor de mummia* should be added to the vitriolic spirit and used as an ointment for the gouty limbs.

In cases of hydropsy half a scruple of *spiritus vitrioli* in *liquore de serapini* should be given in three to four doses a day.

according to the disease. If *liquor serapini* cannot be obtained, *liquor tartari crudi* should be taken instead.

The following procedure should be followed with *pustulae*: for nine days *spiritus vitrioli* should be rubbed into the entire skin. Open sores should be treated by applying *oleum de colcotar* with an alkaline mixture, as is the surgeon's custom. But the bandages must not be changed for six days. Prescriptions and diet should be kept according to the patient's disposition, for the right cure is in medicine only, and not in the food. Therefore, the medicine should be given most assiduously, in order to cure thoroughly the above-mentioned diseases.

Against Sand, Mucus, Grains of Sand and Stones in the Bladder:

Take ten pounds of red colcotar, fourteen half-ounces of *alcohol vini*, put into a vertically or horizontally resting phial and distill for four days; the fire drives a fine red oil into the receptacle. Put just one drop of it into good wine, mornings and evenings.

An Excellent Arcanum Against Falling-Sickness:

Take fifteen pounds of Roman or Hungarian vitriol (which unlike the others do not turn into copper) a half-ounce each *liquoris peonie, camphore, rasure eburis, spodii;* distill through the retort. Take three pounds of distilled liquor, half a pound each *alcoholis vini correcti, aquarum melisse* and *valeriane,* one pound of colcotar; let it go through the retort and then take one pound of this liquor, two pounds of fresh colcotar; distill for twenty-four hours through a horizontally-placed phial. Then you will have *oleum liquorem* and phlegm together; rectify it, then distill its phlegm in *aqua fortis.* When that has been done, extract the liquor from the sand. At last, when you put the compound into the open fire, the red oil appears. *Dosis phlegmatis* always weighs one quint and should be given against the paroxysm. *Dosis liquoris* is one scruple, *dosis olei* three grains. If this disease has lasted from twenty to forty years, you should administer the liquor. The oil should be given those who have been afflicted for fifty years or more.

Other Procedures Against Falling-Sickness:

Take one pound of Roman or Hungarian vitriol; distill its phlegm; put it again into its own colcotar, according to the art; extract it again; repeat this in the fourth degree of the fire. The dose of this medicine is from half a scruple to one third of a drachm, before and after the attack. When the attack has subsided the patients sleep. You should give it while it is still hot. It should be applied before the attack, while it still boils and can stimulate and stir.

Preparation of Vitriol Against Suffocation of the Womb:

Take four half-ounces of vitriol that has been cleared of phlegm and colcotar, six half-ounces of *hertzpoley*, one half-ounce of *alcohol vini*, reduce this compound by distillation; give the same dose as in falling-sickness.

Preparation of Vitriol Against Siphita stricta and Gutta:

Take one half-ounce of the above-described rectified vitriol, four half-ounces of *alcohol vini*, one half-ounce of *alumen jameni* (which should be chalk white and of a sweetish taste mixed with acid); reduce the compound through the fourth grade of the fire in liquor. A dose of six to ten grains of this has to be taken. Apply one half-scruple externally in the spot where the disease takes its root and about which the patient is complaining, that is, *super locum symptomatis*. Symptoms of this disease are in *pulsu oculorum, colli, pulsu ambarum manuum*. Note that those suffering from *syphita stricta* walk in their sleep at night. Tie the medicine over the pulse of their eyes, and if this does not help blindfold the patients with it and cover the pulse of their neck with it, but do not beat them as has been taught by Avicenna.

In Drops or Guttae:

Spread this liquor over the tip of the tongue and keep it wet; if the patient's mouth is shut as in paralysis, open it with an instrument, and put the liquor on his tongue so that it remains wet, because this is the seat of the symptoms and of the disease.

Addition Against Epilepsy and All Kinds of Falling-Sickness:

Take one drachm of prepared vitriol; *liquoris fisci quercini* and *orizontis*, fourteen grains of each; *fiat mixtura*. The seat of falling-sickness is in the region of the nucha. Apply it at the nucha up to the diameter of a penny.

Addition Against Suffocation of the Womb:

Take seven grains of this liquor of vitriol, one drachm *granorum actis, alcohol vini ad pondus omnium, reduc in compositionem, locus suffocationis matricis est in umbilico;* therefore apply it over the navel. If, however, the suffocation appears with vomiting and choking, the medicine should be applied downward.

On Sulphur or Earth Resin

There are so many secret things in sulphur that not enough can be said about it. If it is separated from its arcanum and the dirt is washed off, it becomes whiter than snow.

Sulphur embrionatum from gold is excellent for the heart, that of silver excellent for the brain, from copper for the kidneys, from lead for the spleen, from iron for the gall-bladder, from tin for the liver, from *argentum vivum* for the lungs. But all of these are for one disease; that is, suffocation of the limbs remaining after apoplectic fits, which it is able to suppress and choke. These virtues will be followed by those of *sulphur mineralis*, for they are the same in all works, but they have to be applied with greater strength and more carefully internally and externally. Sapphire has the power of removing *anthraces* and their scurfs, more than all corrosives; its sulphur is of the same kind, for if it is extracted from the *corpus generatum* in the vulcanic preparation, it acts like a plastercast. It does that not only in *anthrax*, but also if applied in cancer, and *in persico igne*. It starts with an *apostema*. There you can see the effects of separation, graduation, and correction. As

to the embrionatic sulphurs in *cachimiis alchamiis opio,* in magnets, *antimonio* and *talc,* you should know that if they are
extracted from bodies and living things, they give extremely
fine sulphur. This property and effect depend on its grade in
the operation. First they take their origin as to their virtues,
even if they have already been spoiled by the preparation.
Secondly, they are best for *phlegmaticis,* especially for *phtysicis,*
peripneumonicis, etc., and in every kind of cough. What nature
can possibly achieve is done by that sulphur and put right again.

Now you know about sulphur and, similar to it, the *spiritus*
vitrioli, as for instance the salts which produce a marvelous
sulphur, in which the *corpora animata* are separated from
the *corpora embrionata;* that is, salt, *sal gemmae, species*
aluminis, vitrioli. I shall, however, give you a short general
rule that all sulphurs from vitriols and salts are *stupefacientia,*
narcotica, anodina, somnifera, but of such quality that the
somniferous effect is brought about in a mild and easy way and
removed again without any bad effects. There is no opiatic
effect as in *hyosciamo, papavere, mandragora,* but it works
mildly and well, without any infection. I have personally prepared and corrected such *somniferum* and *stupefactivum* with
such excellent results. And as we, the physicians, know that the
somnifera do many excellent things, and that the *opiata* contain such poison that they cannot be applied without quintessence, we should all the more rely on and trust in these *somni-*
fera, because we know of many diseases which cannot be cured
without anodines. God has given us a cure for them through
the nature of the anodine.

As for sulphur, you should know that of all kinds that of
vitriol is most widely known; it is firm. Besides, it is so sweet
that chickens eat it and then fall asleep, but wake up again
after some time without any bad effect. You should know that
this sulphur can cure any illness which is to be cured by
anodines, without any bad after effects. It extinguishes pain
and soothes the heat and painful diseases. It is a medicine
preferable in every respect; the cure afterwards is *confortativum*
quinte essentia. What can you physicians do? These two medi-

cines are better than all those of Apollo, Machaon, and Hippocrates. Remember, you physicians, that this sulphur is *sulphur
philosophorum,* because all philosophers have tried to have a
long life and good health and to resist illness. This they found
in this sulphur; therefore, according to their request, it was
called *sulphur philosophorum* and you should know how to
grade, separate, and refine it.

There is still another kind of *sulphur embrionatum,* which is
in wood, and above all, fire itself is sulphur which cannot be
kept in one's life but perishes and dies with the wood. There
is sulphur in all things made of wood or in anything which
burns into ashes. Such sulphur is *vegetabile,* not steady, not
to be used for anything but those things which should be prepared with fire. You should know, however, that such sulphur
indicates the nature of other sulphurs; like fire which devours
everything, each sulphur is an invisible fire which devours diseases, and as the fire does not consume the wood visibly, each
sulphur does it invisibly. Therefore, *oleum ignis* is a great
arcanum in all diseases. And sulphur is *oleum ignis;* if you
wish it to have the effect of a medicine, it must be brought to
its volatility so that it disappears like the flames; it must become
subtle and leave its body; its body must separate from what
is not *elementum ignis.* If sulphur is made subtle and volatile,
then it consumes what should not exist, what is not safe by
nature. Diseases for instance are not steady, but the body is
in opposition to all elements. The *oleum ignis* is only opposed to
things in the body which are not safe; that is, diseases, etc.
Therefore, we distinguish two kinds of *sulphur embrionatum,*
one which is steady, the other which is pure fire. One is a living
fire, the other is an insensible fire. But both the sensible and
the insensible serve for consumption, one in wood, the other
in diseases.

On Mineral Sulphur

Sulphur mineral should not be used in its crude state,
but separated from the impurities. Then it becomes an excellent medicine. If it is elevated from *aloepaticus* and *myrrha*

two to three times, it becomes such a preservative against plague, pleurisy, and all ulcers and constipations of the body that, if taken in the morning, no disease can enter the body that same day, such as plague, pleurisy, and others. This is especially true if it is taken in the following prescription: take ten ounces of well-prepared sulphur (as described), one and a half drachms *myrrhae rubae*, one ounce of *aloehepatici*, one half-ounce *croci orientalis*; mix them and prepare powders from the compound.

If elevated from the vitriol several times—the more the better—it inflames *essentia* and *spiritus vitrioli* in it, and is a preservative against all fevers and a cure for old and new coughs, more than could be disclosed or written in a book. It also is a preservative against falling-sickness during the patient's youth. If taken daily, it will conserve health and protect people so that no harm can afflict them. In trade it corrects wine, so that all wines can be kept fresh and in good condition through it, and wine can be made a healthy drink for man. But it must not be taken in its crude state. If it is in wine, nothing impure can remain in the wine. All wines prepared in this way are of the kind that do not produce sand and tartar, nor do they provoke apoplexy or grave apoplectic fits, coughs, and so on, nor putrefaction; for as we have said, its arcanum is of such quality, when prepared as has been described, that nothing can be found to equal it. Therefore, watch mineral sulphur; it should not be prepared only once, but often and repeatedly, so that it loses all its impurities, poison and what should not be in it; only that which contains the best medicine should remain. The right kind of sulphur has the quality of making red things white by its fume, as for instance red roses.

We also know that it turns white if refined and applied in medicine as described; but it must be applied externally. It should be known that there are many colored sulphurs: yellow, dark yellow, red, redder, brown, black, white, green. But all these colors are of no use, except the yellow one; the more yellow, more gold the color is, the better for health. The others contain much of *arsenicum realgaris* and similar things; therefore they should be avoided in medicine. But in alchemy these others are better because of the " impression " which they have received from such " realgaric " spirits.

Sulphur also removes external ringworm of the body, and for this the less colored sulphurs are better than the yellow-red ones, because of the subtlety of the arsenic spirits. If such sulphurs are sublimated from vitriol, *sale nitro, sale gemme,* or *alumine plumoso* for several times, then they are so fine that they remove ringworm and *serpiginem* from the base and with the root. It is a great help in removing externally what has originated internally, bringing it out from inside and healing it. There are magnetic powers and ways which cannot be sufficiently explained, except by the great experience of the " vulcanic art," for thanks to it there are great and wonderful miracles in nature and its powers. Therefore, remember that sulphur has such a nature and quality that it gives man excellent health, if it has been graded; this is true not only if it is swallowed, but also its smoke preserves and conserves if *Meisterwurzen, Krammetwein,* and rosmarine are added.

On the Use of Metallic Sulphurs

The sulphurs which are made out of whole metal, the primary sulphurs, are separated by alchemic means in the ways I have described with regard to sulphur. The same are also in these six metallic sulphurs. They possess even more physical properties and special qualities, because they have turned into metal. The properties of the metal are to found in the sulphur. Therefore it follows that these are much better and more noble than other sulphurs. The physician should know that virtues contained in sulphur are contained in this one especially, and graded to the highest. Therefore, the sulphur coming from gold has gold-like virtues, that from silver silver-like virtues; that from iron has the nature of iron, and does what iron does. What is done by topaz and *crocus martis* is also done by the sulphur which comes from iron. It is the same with copper, *saturno,* and other metals, and the physicians should do their

best to have such sulphur, for though the dose be small, the
effect is great so that even lepers recover.

On the Alchemistic Virtues of Sulphur
First, On the Embrionic Sulphur

The extraction of embrionic sulphur is sometimes performed
by sublimation, sometimes *per descensum,* if it is well-disposed
sulphur and not much mixed with other bodies. Sometimes,
however, if it is very subtle, it cannot be sublimated by simple
distillation, but only by *aquae fortes.* It may be extracted
from the other bodies into the *aquae fortes* and afterwards it
coagulates. There are several kinds of *aquae fortes,* and they
do not take away or change anything in the shape of the sulphur
or in its power, if they are extracted in the right way, according to their properties. Afterwards it becomes fixed, if it contains gold; usually gold is found by sublimation. It also fixes
all volatile sorts of gold and holds them, which cannot otherwise be done. It is not noticeable during the separation because
of its subtlety and its subtle " corporality " (*corporalitet*).
Many procedures have been started with such sulphurs, such
as making tinctures from them, but the formula could not be
found, for there is not enough in sulphur in which to hide a
tincture. Therefore, such searching is in vain, unless it contains gold; otherwise one should not look for silver in sulphur;
none of them contains silver. In one there is more gold than in
the others, as for instance in red talc, marcasite and goldsand,
Those are rarely without gold. But those who want to attempt
it, should remember that the sulphur must be separated from
the gold in such a subtle way that the gold does not evaporate.
I could help many now with a few words, but I prefer to keep
silent.

On Mineral Sulphur

Now I must write about the miracles of the qualities of
sulphur in alchemy. Many have tried various arts, trying to
make something from sulphur which would be more than sul-

phur itself. But God has created art in such a way that it can do it. As the power of art can do it, the master has followed that art and tried to find out what can be made of sulphur which is not in sulphur or has tried to find out if it might lead to something else. It is like a woman who in herself cannot bring forth children, but who, together with her husband, does. If art can bear a little more and if they do something together, then the artist is the man and father who causes all. Now art has found a prescription for *spiritus transmutationis,* to make a liver or lung from linseed oil or sulphur. Such a liver and lung have been frequently distilled. While working at it, it was found that such a liver gives milk which does not differ from ordinary milk. It is fat and thick; also such a liver gives red oil like blood, so that milk and blood were distilled together. Neither colored nor falsified the other, but each separated from the other without doing any harm—the white as a sediment, the red on the surface.

Now the art went on searching, trying to make silver from the white and gold from the red. I know that nothing has ever been made from the crude or milk sulphur, either by the ancients or by the moderns. Therefore, I say that this is a dead milk with nothing in it. But mark what I say about the red oil which is yielded from this liver. Each well-polished crystal or pearl which is put into it for some time—for three years—becomes a hyacinth, that is, like a hyacinth in all its character and looks. Likewise a ruby which is not highly graded and which is put into the red oil for nine years becomes so pure and clear and reaches such a grade that, if later it is put into darkness, it gives light like a coal, so that you can see from all around where it is lying. This has been proved by experience. The old alchemists made carbuncle out of it; if they put a good hyacinth into the oil, it was to become a burning carbuncle. It is a matter of experience that such color is not only in the one I have described. For it also turns sapphire light blue with green in it, and it colors other precious stones. But it colors no other things, not even glass. It grades the precious stones so highly that they reach the highest quality, higher than nature itself can grade them. Such gradation of precious stones has never been described or used before; it is only possible with those colored and tinted by the red blood in sulphur.

It should also be known that every kind of silver put into it and left to lie there for a time turns black and gives a chalk sediment, not steady and fixed, but unsteady and volatile. If it has been there all its term, and the end comes, it does all that can be done; it is of no use to speak more of it. Remember that sulphur if brought to the right grades becomes more subtle, finer, stronger, quicker in its effect, and better. This is the effect of the tincture on stones and metals. Those who want to make it should not just believe that they know it, but should really know it, for its preparation is the most dangerous work in alchemy. It must be thoroughly tried and frequently used, and must be known not only by hearsay but by one's own knowledge and experiment. This is twice as good. I cannot say anything about its virtues, and whether they are graded or not. I can only speak about color and tincture which, as I have said, are of the highest. But mind that this will not happen in the virtues and their powers, for there are no *tinctura virtutis,* but only *coloris.*

On the Sulphur of Metals to be Used in Alchemy

I have spoken several times in this chapter about the oil of sulphur, which is made from metals by destroying them. I also indicated what properties they have for medicine, that tinctures can be made from them, and that one can be tinctured into the other; but I was not able to go into this deeply. I need not give the motive for it here. But it is really true that those who have *sulphur auri* are able to grade gold with it above the normal degrees, that is, above 24 and up to 36 degrees and more, so that the gold color could not come higher or be more permanent in *antimonio* or *quarrierung.*

It is further true that *sulphur lunae* grades silver so highly in its whiteness that on a needle equal quantities of silver and copper cannot be distinguished; all is pure, fine, clean silver.

You should also know that sulphur from copper can stand the thunderbolt, but does not become graded, and its color mixed with sulphur of iron becomes the best steel; with *sulphur iovis* you obtain the best of all tins, which can even resist the thunderbolt; by means of *sulphur saturni* lead becomes firm, so that it yields no white lead, yellow lead, *minium,* nor crystal glass, nor any spiritus. The sulphur from *argentum vivum* makes it possible to work at *argentum vivum* with the hammer; it can be treated with red-hot, iron-like copper, but it cannot bear the *cineritium.*

So much for metallic sulphurs; also, if you cast *sulphur auri* into silver it colors it, but fixes nothing. In this way you can transform sulphur into another substance than itself; but not into what you would like. This is what you should know of sulphur: its kinds, its nature, qualities and character. For handling sulphur a good chemist, a perfect artist, an experienced master, a thorough experimentalist is necessary.

IV

A BOOK ON NYMPHS, SYLPHS, PYGMIES, AND SALAMANDERS, AND ON THE OTHER SPIRITS

BY

THEOPHRASTUS VON HOHENHEIM

CALLED PARACELSUS

TRANSLATED FROM THE GERMAN, WITH AN INTRODUCTION

BY

HENRY E. SIGERIST

INTRODUCTION

We do not know exactly at what time Paracelsus wrote his treatise *On Nymphs, Sylphs, Pygmies and Salamanders*. It must have been in his later years when he wrote other philosophical and theological books, and Sudhoff published it in the last volume of his edition [1] together with such treatises as *De sagis et earum operibus, Liber de lunaticis, De generatione stultorum* and others.

It is a very unpretentious treatise, compared with such books as the *Philosophia sagax*. At times it is written in a style that reminds us of fairy tales. And yet it gives an excellent picture of Paracelsus' philosophical and theological views.

Paracelsus was a man of the Renaissance and a rebel, not only in medical matters. His intense individualism made it impossible for him to accept the integral dogmas of the Catholic Church and its scholastic philosophy, just as he was unable to follow the doctrines of the Reformation. He disliked the spectacular cult of the Church and had a deeply felt longing for the simplicity of the early Christian community. The Bible was the starting point of his thought.

God is almighty. In the beginning he created the heaven and the earth and all that is in it. He made man from the dust of the ground, from limus terrae, and made him in his own image. Limus terrae, according to Paracelsus, was an abstract from heaven and earth, the quintessence of the four elements and the firmament; and since man was made from it he is a microcosm, reflecting the macrocosm and intimately connected with it. Heaven and earth, the firmament and the four elements, constitute the world, and also man. Man, therefore, has a dual body, an elemental and a sidereal one. The elemental body, the earthly part of man has material needs. It craves food and drives man to reproduce himself, while the sidereal body, the heavenly part has appetites of another kind. It drives man to

[1] Vol. XIV, pp. 115-151. In the edition of Huser, the text is found in vol. IX, pp. 45-78.

seek wisdom and satisfaction in the cultivation of the arts and sciences. This duality explains many tensions and conflicts. Man, however, was created in the image of God. God gave him a soul. That made him a moral being that seeks salvation in God. These three parts, these three bodies of man are one, an indivisible whole, as long as he is alive, but they separate in death, each returning to its origin.[2]

The destiny of man is to walk in the light of nature and in the eternal light. " The beginning of theology is the enlightenment of man. He must be enlightened in all things of nature and must know them." [3] If he does this, he will understand Christ, and the Holy Scriptures will not be dead letters to him. Theology, therefore, requires philosophy which, to Paracelsus, is nothing else but the exploration of nature. Philosophy and science are identical to him. God has created the world and reveals himself in his work. Nature can be explored and can be interpreted. God made it thus that the human mind can perceive its light. " Nature emits a light by which it may be perceived from its own radiance." [4] And through the light of nature man gets access to the eternal light, to God's own world.

The investigation and interpretation of nature thus becomes the subject of philosophy and the door to theology. The interpretation of nature, however, is not easy and sometimes presents great difficulties. There are strange things in nature that seem incredible and about which nothing can be found in the Holy Scriptures. But God is almighty and can create whatever he pleases, strange as it may appear to man. And the purpose of the Holy Scriptures was not to describe and explain every object of nature; they deal with God and the soul in their interrelationship.

Such strange creatures are the mysterious beings—Paracelsus usually calls them *ding,* things—that inhabit the four

[2] See B. S. von Waltershausen, *Paracelsus am Eingang der deutschen Bildungsgeschichte,* Leipzig, 1936, a book that gives by far the best interpretation of Paracelsus' philosophy and theology.

[3] See p. 247.
[4] See p. 223.

elements, the nymphs, sylphs, pygmies and salamanders and, related to them, the sirens, giants and dwarfs.

To the mind of primitive man, living in constant intercourse with a still hostile nature, every object of his environment seemed animate. The rock, the tree, the pond, all were inhabited by spirits from whom good or evil might come, who forced one to be on his guard constantly and who had to be placated at times. The creative genius of the Greeks lent color to these vague spirits. The heaven became Zeus, the sun Apollo, the moon Artemis, not mere forces of nature, but beings with a biography, with passions and whims very similar to those of mortal man, and thus very close to him. And, in his immediate surroundings, springs and brooks were populated with lovely creatures, the nymphs, some of whom, like Calypso, achieved great fame. Dryads spoke from the trees, and on hot summer afternoons Pan's pipe was heard in the woods. Sirens half woman, half bird, lured the traveller on distant seas and there were tales of far-away regions inhabited by one-eyed giants, the Cyclops. In the vicinity of mountains Vulcan's and his assistants' blows could be heard as they rocked the earth with their hammers. From Homer to Ovid and later the doings of these deities were pictured in endless numbers of songs. They were real to the naive man and presented to him a humanized picture of a nature that he could understand and with which he could deal. We still feel sentimentally attached to this colorful world of Greek mythology and it still lives in our language. We speak of venereal diseases, of nymphomania. We vulcanize rubber, use salamanders in the kitchen, call small races pygmies and know sylph-like creatures.

Christianity had a difficult task in overcoming the ancient gods. Its monotheism was too austere for the common man, and soon the Christian heaven was populated with angels and saints, many of whom had taken over the functions and attributes of ancient gods. Venus took abode in the Harz mountain and continued to seduce men, promising them the joys of love and eternal youth. And when she could not act so openly, she was satisfied in being worshipped under the features of Saint

Veronica. Her bird, the dove, became the Holy Ghost from whom Mary conceived without sin.

Christianity succeeded in dethroning the major gods of paganism or in filling their place with its own saints, but it had no substitute to offer for the elemental spirits, for all those minor deities and demons that were so close to man in his everyday life. There had to be an explanation for good luck and bad luck. Omina would explain a great deal and the belief in signs persisted to our days. But if you dropped a piece of buttered bread, why was it that it usually fell to the ground with the buttered side down? Was it not logical to assume that this was a practical joke of some goblin?

And so, the ancient spirits of nature survived the twilight of paganism. Greek demons joined the legions of spirits of the Germanic and the particularly rich Celtic mythology. As heretofore, man was surrounded by a world of strange beings whose dealings with men were recorded in hundreds of fairy tales.

The Church could not deny their existence; they were too deeply rooted in popular belief. But the theologians—not the true theologians, as Paracelsus said—declared them to be devils that had to be fought. Paracelsus had totally different views. He too believed firmly in the existence of such beings and considered them important enough to devote a special treatise to them. He never says that he had seen them. Like angels they appeared only rarely to man. But on his peregrinations he heard a great deal about them. In the mountains he saw strange excavations which local traditions attributed to the activity of pygmies. And in certain regions, at night, he probably saw flickering lights running around over the ground playfully, will-o'-the-wisps, and people said they were the spirits of fire.

Paracelsus was particularly interested in the elemental spirits. Four elements constitute the material world: water, air, earth, and fire; and each element is inhabited by a kind of being peculiar to it and to whom the element is his *chaos*. Man lives between heaven and earth. Between heaven and earth there is air. Man, therefore, lives in air, and air is his chaos. Thus water is chaos to the nymphs, earth to the pygmies, fire to the salamanders, while the sylphs have the same chaos as man. They are at home in their chaos and, therefore, nymphs do not drown in water, pygmies are not choked in earth and salamanders do not burn in fire. This seems incredible but God is almighty. Why should he not be able to create such beings?

They are not devils but creatures of God and like other creatures they are testimonies of God's miraculous works. They are not spirits and it is a mistake to call them elemental spirits, for they have bodies that resemble that of man. They have two of man's bodies, the elemental and the sidereal ones. This explains why they too eat food and feel the urge to reproduce themselves. And it also explains why they are skillful in the arts and crafts. And since they have these two bodies they are, like man, subject to the diseases that arise from the elements and from the heaven. But they lack one of man's three bodies, and this makes them beings of their own. They have not the soul. In that they are like animals. They look like men but are not descendants of Adam. They live in social groups, obeying laws, bear children but when they die, there is nothing left.

Since every creature that is denied salvation craves it, they seek union with man. When a nymph marries a man from Adam, through the sacrament of matrimony she receives the soul. And her children have the soul. She becomes a moral, responsible being who will have to render accounts on the Day of Judgment. Tradition reports of many such nymphs who married man, lived with them and were faithful to them.

Since they are not devils but beings that may be saved, they must be treated kindly, and contracts and agreements with them are binding for man too. Paracelsus sides passionately with the nymph who married a nobleman, von Staufenberg, and was repudiated by him because he took her for a devil. He married another wife but the nymph came back and killed him on the day of his wedding. Paracelsus finds that she was justified in doing it, since no judge would have given her her rights as she was not from Adam, and so God put the law into her hands and allowed her to punish the adulterer.[5]

[5] See p. 245.

For what purpose did God create such strange beings? As was mentioned before, they are testimonies of the grandeur of God's works. This they share with the whole of nature, but God created them for a special purpose—and here Paracelsus is writing as a theologian and scientist. God created these elemental beings as makers and guardians of the treasures of the earth. There is an infinite wealth of minerals in the earth. They are made in the depths of mountains under the influence of fire, and this is where the salamanders come in. Once the mineral ores are made they are guarded, those in the earth by the pygmies, those on the surface by the sylphs, and those at the bottom of the waters by the nymphs. Why should they be guarded? Has God not permitted these treasures to be made for man? Yes, but they should not be discovered all at once. God decides when the time has come for a deposit to be found and exploited. Then the pygmies, or whoever the guardians are, retire and the place is open to man.

Paracelsus, who on his wanderings visited many mines and worked in them, must have heard folk-tales that inspired him to propound such an interpretation. Mines were always considered places of a special kind. There could not be such spots of concentrated wealth without some reason, and was it not logical to assume that the earth would not yield her treasures without resistance? Many writers on mining mention the spirits that inhabit the mines and threaten the workers' lives, and many superstitions connected with mining have survived to our day. Paracelsus frequently heard of local traditions according to which a mining region had been inhabited by strange beings in the days of old.

The elemental beings have one more purpose. They are a warning to man. Their appearance, like that of angels, reminds man of God's almightiness and is a warning for him to amend his life and to serve God in humility and devotion. They also warn him that God could exterminate the human race and still keep the earth populated with beings similar to man.

Toward the end of the treatise Paracelsus describes another kind of beings still more strange than the former, the sirens, giants, dwarfs and will-o'-the-wisps. He has a scientific explanation for their origin. They are monsters. Just as two perfectly normal human parents can give birth to a monster, so do the elemental beings. Thus, the sirens are monsters of the water people, the giants of the sylphs, the dwarfs of the pygmies, and the will-o'-the-wisps of the salamanders. Like human monsters they occur rarely and their kind dies out soon. They are abnormal happenings like comets and earthquakes, and their appearance signifies impending disaster. Thus they are powerful omina, like comets and earthquakes.

Paracelsus collected the various popular traditions about the elemental and other spirits that had come down from pagan antiquity. He did not render them the way he found them, and did not accept the official view that they were devils. He pondered over them and tried to determine their place and function in the system of nature. Since personal observations were hardly available to him, he had to rely on traditions; but he analyzed them as he did other phenomena of nature, other creations of God. And he found a place for them in nature where their existence and function had meaning and significance. He did it without having recourse to the devil for an explanation, and his attitude toward them was infinitely more humane than that of most of his contemporaries.

In this as in other treatises Paracelsus expressed views that were not orthodox. In this book he opposed not the physicians but the theologians, and he did not expect to have his views accepted during his lifetime. He wrote such books because he could not help writing them, in order to clarify his own thought, because he wanted to understand heaven and earth, God and nature. The last paragraph of the treatise is at the same time bitter and hopeful. Paracelsus feels misunderstood, but the day will come when the real and the false scholars will be revealed for what they were, those who wrote the truth and those who lied, those who were right and those who were wrong. At that time he hopes that his work will find due recognition.

Like the other theological writings, the treatise *On Nymphs, Sylphs, Pygmies, and Salamanders* was not published during

Paracelsus' life. It was first printed in 1566 as a separate treatise.[6] In 1567 it was included in a collection of his philosophical writings [7] which was later translated into Latin and published in 1569.[8] The treatise appeared again in Huser's edition of the Collected Works, having been copied, as Huser said, from Paracelsus' own manuscript.[9] In the early 19th century it was again printed, twice in German and once in Italian.[10]

Rationalism killed the belief in elemental spirits, but they continued to live in fairy tales and remained a favorite subject of poets and writers who made frequent use of Paracelsus' treatise. In our days, Jean Giraudoux, in the preface to his charming play *Ondine*, has acknowledged his debt to Paracelsus.

A history of the elemental spirits in religion, theology, philosophy, literature and art has still to be written. It is a fascinating topic, and its treatment would illustrate the history of civilization from many angles. In such a history Paracelsus would take an important place. His treatise not only presents the subject in a new light but also reveals the personality of its author, his approach to nature as a scientist and mystic. It also is a testimony of his profound respect and deep love for all objects of nature. With Leonardo da Vinci he could have said:

L'amore è tanto più fervente, quanto la cognizione è più certa.

[6] *Ex libro de nymphis, sylvanis, pygmaeis, salamandris, et gigantibus etc.* Nissae Silesiorum, Excudebat Ioannes Cruciger, 1566.

[7] *Philosophiae magnae Tractatus aliquot.* Getruckt zu Cöln, bey Arnoldi Byrckmans Erben, 1567.

[8] *Philosophiae magnae Collectanea quaedam.* Per Gerardum Dorn è Germanico sermone . . . Latinè reddita. Basileae, apud Petrum Pernam, 1569.

[9] Neundter Theil *Der Bücher und Schrifften* . . . Paracelsi . . . Durch Iohannem Huserum. Getruckt zu Basel, durch Conrad Waldkirch, 1590, pp. 45-78.

[10] In: D. Valentin Schmidt, *Beiträge zur Geschichte der romantischen Poesie*, Berlin, 1818.—*Blätter für höhere Wahrheit. Aus ältern und neuern Handschriften und seltenen Büchern. Mit besonderer Rücksicht auf Magnetismus.* Herausgegeben von Johann Friedrich von Meyer. Zweyte Sammlung. Nebst einer Abbildung in Steindruck. Frankfurt am Mayn, bey Heinrich Ludwig Brönner, 1820.—*Ondina*, Racconto Del Barone Federico De La Motte Fouqué con un estratto di Teofrasto Paracelso sugli esseri elementari. Milano, Per Ant. Fort. Stella e Figli, 1836.

LIBER

DE NYMPHIS, SYLPHIS, PYGMAEIS ET SALA-
MANDRIS ET DE CAETERIS SPIRITIBUS
THEOPHRASTI HOHENHEIMENSIS

Prologus

We are familiar and well acquainted with the creation of all
and every natural object that God has brought into existence.
Every country has cognizance of its own, of what occurs and
grows in it. Every man has cognizance of himself. Thus also
an art and a craft have knowledge of their subjects. In such a
way cognizance is obtained of all creations that God has
brought into existence, and nothing is concealed that man did
not or would not know. That does not mean that all knowledge is in one, namely, that one man knew all this, but each
one his field. When they all come together, then everything
is known. It does not mean either that every country has
knowledge of every other country, but each of its own, and
when they are all taken together, then all is known. And every
city, village, house, etc. has its own cognizance of all natural
objects. And in the arts and crafts, all created objects are being
used, one in this, one in that, and while they are all being used,
we come to understand for what purpose they have been created.
And the final conclusion is that everything is subservient
and subject to man. There are more things, however, than those
which are comprehended and recognized in the light of nature,
things which are above and superior to it. But these cannot
be understood against the light of nature, that is in the light
of nature, but in the light of man which is above the light of
nature. There they may be understood. For nature emits
a light by which it may be perceived from its own radiance.
But in man there is also a light, outside the light born in
nature. This is the light by which man hears, learns and penetrates supernatural things. Those who search in the light of
nature, talk of nature. Those who search in the light of man,

talk of more than nature. For man is more than nature. He is nature. He is also a spirit. He is also an angel. He has the qualities of all these three. When he walks in nature, he serves nature. When he walks in the spirit, he serves the spirit. When he walks in the angel, he serves as an angel. The first is given the body, the others are given the soul and are its treasure. Since man has a soul and the other two, he elevates himself above nature, to explore also that which is not in nature, especially to learn about and explore hell, the devil and his realm. In the same way man also explores the heaven and its essence, namely God and his realm. For he who must go to a place should know the character of the place in advance; then he will be able to travel wherever he pleases. Know therefore that the purpose of this book is to describe the creatures that are outside the cognizance of the light of nature, how they are to be understood, what marvellous works God has created. For it is man's function to learn about things and not to be blind about them. He has been equipped to talk about the marvellous works of God and to present them. It is possible for man to explore the essence and qualities of every single work that God has created. For nothing has been created that man could not explore, and it has been created so that man may not be idle but walk in the path of God, that is, in his works. Not in vice, not in fornication, not in gambling, not in drinking, not in plundering, not in the acquisition of goods nor in the accumulation of riches for the worms. But to apply his spirit, his light, his angelic kind to the contemplation of divine objects. There is more bliss in describing the nymphs than in describing medals. There is more bliss in describing the origin of the giants than in describing court etiquettes. There is more bliss in describing Melusine than in describing cavalry and artillery. There is more bliss in describing the mountain people underground than in describing fencing and service to ladies. For in these things the spirit is used to move in divine works, while in the other things the spirit is used to seek the world's manner and applause, in vanity and dishonesty. He who experiences and hears much on earth will also be learned in the resurrection while he who knows nothing, will be inferior. For there are many mansions in God's house and each one will find his mansion according to his learning. We are all learned but not equally, all wise but not equally, all skillful but not equally; he who searches deeply is most. For research and seeking experience thrive in God and avoid the vices of the world; they flee worldy service, obedience to princes, court manners, nice gestures; they learn tongues, in which are also lies and curses. But God's miraculous works—man's light learns them and does not ask tongues for it. Obedience to God, this is man's command. Obedience to men, what else is it than a vain shadow? Man does not pay for obedience, has no reward for it. He dies, and when he is dead, he is dirt. What does man make of himself? He must learn more than obedience, and let obedience be, and love his neighbor; then obedience will come out by itself, as the flower and fruit come out of a good tree. Oh, how joyful is he who thinks in the following of his creator! He finds pearls that will not be thrown to the swine. But he who thinks in the following of man searches for pearls like a sow which stirs up everything and does not find anything of use. And so, know how to understand the beginning of this book. I am not writing about lovely matters and nice stories, but about supernatural matters that do not require nice stories, but let gossip be what it is.

LIBER INCIPIT THEOPHRASTI

TRACTATUS I

CAPUT I

To begin with, it is only fair that I explain the subject about which I write hereafter, what it is all about. Know, therefore, that the purpose of this book is to describe the four kinds of spirit-men, namely, the water people, the mountain people, the fire people and the wind people. Included among these four kinds are the giants, the melusines, the Venusberg and what is similar to them. We consider them to be men, although not from Adam, but other creatures, apart from man and all other animals, in spite of the fact that they come among us and children are born from them, although not of their own kind, but of our kind. How things are in relation to us will have to be described in the following order: First, their creation and what they are; second, their country and habitation, where they stay and what their mode of living is; third, how they come to us and let themselves be seen, how they mix and have intercourse with us; fourth, how they perform some miraculous works, such as Melusine, the Venusberg and similar stories; fifth, the birth of the giants and their origin, and also their vanishing and return.

It is true that a philosopher must build on the Holy Scriptures and take his arguments from them. But in the Scriptures nothing special is written about these things, what to make of them or how to explore them. There are some remarks about the giants only. Although these things are treated outside the Scriptures, their exploration is justified by the fact that they appear and exist. These things must be explored just like magic, if we are to believe in it—and we do and wish to ascertain its origin. The writers of the Bible and the New Testament discussed how the soul has to act toward God and God toward the soul, so that this philosophy may not fall back. But if we are permitted knowledge about the devil and about spirits and similar subjects, then this is something that has to be explored also, in regard to its nature. We have the power to travel in all the works of God, the physician in the natural remedy, the apostle in the apostolic remedy. For, as a sick man calls for a physician, things call for a philosopher, and a Christian for his Redeemer, and every work for his master. Such creatures are also necessary and also represent their condition, and they have not been created in vain. For, Samson was a man. Now, he had strength over all men, which was not natural nor credible, and it resided in the hair; this is considered by men as superfluous and unnecessary, but it has its reason why it must exist. David who was a small man, killed the giant Goliath. It had to be, and yet did not seem humanly possible. Hence, nothing has been created that has not a mystery, and it may be a great one. In the Old Testament queer stories happen which nobody can interpret, and then the New Testament teaches him about them; how things are finally found in a given place, why they occurred and from what cause. Thus one learns that they occurred indeed. The same applies to the things about which I write in the following—people find them unnecessary and find that it is useless to talk about them. But since there is a reason for them which will be apparent, about which I shall write the sixth treatise, therefore it is not unnecessary but necessary to explore the things, and to include them in our philosophy and speculation, to contemplate these things together with others, what their final purpose is.

CAPUT SECUNDUM

SPIRITUS QUID ET ANIMA; ITEM SPIRITUS HORUM CARO EST ET CARO SPIRITUS; EXEMPLUM RESURRECTIONIS.

The flesh must be understood thus that there are two: the flesh from Adam and that which is not from Adam. The flesh from Adam is a coarse flesh, for it is earthen and is nothing else but a flesh that can be bound and grasped like wood or a stone. The other flesh which is not from Adam, is a subtile flesh and can not be bound nor grasped, for it is not made from earth. Now, the flesh from Adam is the man from Adam. He is coarse like the earth which is compact. And, therefore, man cannot go through a wall. He must make a hole to slip through, because nothing recedes before him. But before the flesh that is not from Adam, walls recede, which means that such flesh does not require doors nor holes; it goes through intact walls and does not break anything. They are both flesh, blood, bone and so on, whatever belongs to a man, and in their whole nature they are like man's. But they are different in that they have a double origin like two cousins; they are in equal way like a spirit and like man. The spirit goes through all walls and nothing locks him out; man, however, not, for he is locked out by the bolt or the lock. Just as you can distinguish and differentiate between a spirit and a man, so you shall recognize the people about whom I am writing but with the difference that they are separated from the spirits by having blood and flesh and bones. With all this, they bear children, talk and eat, drink and walk—things that spirits do not. Hence, they are like the spirits in speed, like man in gestures, figure and eating, and so they are people who have the character of spirits and also that of man, and both are one in them.

Although they are both spirit and man, yet they are neither one nor the other. They cannot be men, since they are spirit-like in their behaviour. They cannot be spirits, since they eat and drink, have blood and flesh. Therefore, they are a creation of their own, outside the two, but of the kind of both, a mixture of both, like a composite remedy of two substances which is sour and sweet, and yet does not seem it, or two colors mixed together which become one and yet are two. It must be understood further that although they are spirit and man, yet they are neither. Man has a soul, the spirit not. The spirit has no soul, but man has one. This creature, however, is both, but has no soul, and yet is not identical with a spirit. For, the spirit does not die, but this creature dies. And so it is not like man, it has not the soul; it is a beast, yet higher than a beast. It dies like a beast and the animal body has no soul either, only man. That is why it is a beast. But they talk, laugh like man;

this is why they are more like men than like beasts, but are neither man nor beast. They are to man like a monkey which is the animal resembling man most in gestures and actions. And as a pig has man's anatomy, being inside like man, yet a pig and not a man—in the same way these creatures are to man as monkeys and pigs, and yet better than they. For they are in everything like men, but without soul, and better than man, for they are like the spirits which nobody can lift. Christ died and was born for those who have a soul, that is who are from Adam, and not for those who are not from Adam, for they are men but have no soul. So much can be gathered from the Scriptures about them that they must be admitted to be men; but as far as the soul is concerned there is no knowledge of their having one.

Nobody should wonder that there are such creatures. For God is miraculous in his works which he often lets appear miraculously. For these things are not daily before our eyes but very rarely; and we see them only in order that we may know of their existence, for they exist, and yet appear to us as in a dream. The great wisdom of God cannot be fathomed, nor can his great miraculous works be fathomed, not more than is needed to recognize our creator in his miraculous deeds.

Now, they are separated from us because they are not from Adam and do not participate in the same earth from which Adam was made, but God has decreed that we may see them miraculously, which has a particular meaning, as will be discussed in the last treatise. They have children and their children are their kind, not ours. They are witty, rich, clever, poor, dumb like we who are from Adam. They resemble us in every way. Just as one says: man is the image of God, that is, he has been made after his image—in the same way one can also say: these people are the image of man and made after his image. Now, man is not God although he is made like him, but only as an image. The same here: they are not men because they are made after his image, but remain the same creatures as they have been created, just as man remains the same as God has created him. Thus he wants every creature to remain at the place where it has been created. And as man cannot boast that he is God, but a creation of God, thus made by God, and God wants it thus—in the same way these people cannot boast that they have a soul like man, although they look like him. Just as man cannot boast that he is God although he is made after him and exists. Thus man lacks that he is not God. And the wild people lack the soul. Therefore they cannot say that they are men. And so the one lacks God, the other the soul. Thus God alone is God, and man alone is man.

And so they are men and people, die with the beasts, walk with the spirits, eat and drink with men. That is: like the beasts they die, so that nothing is left; and neither water nor fire harms them, like the spirits, and nobody can lock them up, like the spirits, but they reproduce like men and therewith share his nature. They have man's diseases and his health but their medicine is not from the earth from which man is made, but from the element in which they live. They die like men, but are dead like the beasts. Their flesh rots like other flesh and their bones like other men's bones and nothing remains of it. Their customs and behaviour are human, as is their way of talking, with all virtues, better and coarser, more subtile and rougher. The same applies to their figures: they are very different, like men. In food they are like men, eat and enjoy the product of their labor, spin and weave their own clothing. They know how to make use of things, have wisdom to govern, justice to preserve and protect. For although they are beasts, they have all the reason of man, except the soul. Therefore, they have not the judgment to serve God, to walk in his path, for they have not the soul. The beast from inborn nature seeks a just course toward itself, and so do they, but they have the highest reason, above all other animals. Just as man, above all creatures, is closest to God on earth in intelligence and faculties, so they are, among all animals, closest to man and so close that they are called people and men and are held and taken for such, so that there is no difference except in their spirit-like way and in the lack of soul—a queer and marvellous creature, to be considered above all others.

TRACTATUS II

ABOUT THEIR ABODE

Their abode is of four kinds, namely, according to the four elements: one in the water, one in the air, one in the earth, one in the fire. Those in the water are nymphs, those in the air are sylphs, those in the earth are pygmies, those in the fire salamanders. These are not good names, but I use them nevertheless. The names have been given them by people who did not understand them. But since they designate the things and since they can be recognized by the names, I shall leave it at that. The name of the water people is also undina, and of the air people sylvestres, and of the mountain people gnomi, and of the fire people vulcani rather than salamandri. Whatever it may be, and however the differentiation may be understood, let it stay. Now you must know that if their regions have to be described, they must be divided into their parts. For the water people have no intercourse with the mountain people, nor the mountain people with them, nor the salamanders. Each has his special abode, but they appear to man, as mentioned, so that he may recognize and see how marvellous God is in his works, that he does not leave any element void and empty, without having great wonders in them. And now follow the four regions, which explains the difference of abode, also of person, essence and kind, how far they differ from each, yet more similar to man than to each other, and yet all men, as was explained in the first treatise.

You know that there are four elements: air, water, earth and fire; and you also know that we, men from Adam, stand and walk in air and are surrounded by it as a fish by water, and we can just as little be without it as a fish without water. As the fish has its abode in water, where water takes for it the place of the air in which it lives, so air takes for man the place of water, in relation to the fish. Thus everything has been created in its element, to walk therein. From this example you understand

that the undinae have their abode in water, and the water is
given to them as to us the air, and just as we are astonished
that they should live in water, they are astonished about our
being in the air. The same applies to the gnomi in the mountains: the earth is their air and is their chaos. For everything
lives in chaos, that is: everything has its abode in chaos, walks
and stands therein. Now, the earth is not more than mere
chaos to the mountain manikins. For they walk through solid
walls, through rocks and stones, like a spirit; this is why these
things are all mere chaos to them, that is, nothing. That
amounts to: as little as we are hampered by the air, as little
are they hampered by the mountain, by earth and rocks. And
as it is easy for us to walk through air and air cannot stop us,
so rocks and cliffs are easy to them. And so, things are all
chaos to them which are not chaos to us. For a wall, a
partition, stops us so that we cannot go through, but to them it
is a chaos. That is why they walk through it; to them it is
their air in which they live and walk, as man does in the air
that is between heaven and earth. And the coarser the chaos,
the more subtile is the creature; and the more subtile the
chaos the coarser the creature. The mountain people have a
coarse chaos; therefore, they must be the more subtile; and man
has a subtile chaos; therefore, he is all the coarser. And thus
there are different kinds of chaos, and the inhabitants are
adapted in nature and quality to live in them.

Thus one wonder is explained, that of their abode, and you
know now that their habitation in the four elements is their
chaos, just as the air is for us, and there cannot be such accidents to them as drowning, or suffocating or burning, for the
elements are nothing but air to the creatures who live in them.
Since water is the fish's air, the fish does not drown, and so the
unda does not drown either. As in the water, so in the earth:
the earth is air to the gnomi; hence they do not suffocate. They
do not require our air, we do not theirs. Thus also with the
salamanders: fire is their air, as our air our air is. And the
sylvestres are closest to us, for they too maintain themselves
in our air, and they are exposed to the same kind of death as

we, namely: they burn in fire, and we too; they drown in water, and we too; they suffocate in the earth, and we too. For, each remains healthy in his chaos; in the others he dies.

Therefore, you must not be astonished about things that seem incredible to us; to God everything is possible. He has created all things for us, not according to our thoughts and intelligence, but above our thoughts and intelligence. For, he wants to be looked upon as a God who is marvellous in his creations. Had nothing been created but merely what is possible for man to believe, God would be all too weak, and man would be his equal. This is why he has created things as a God, and let man marvel about them, and let his works be so big, that no one can marvel enough about these beings also. God wants to have it thus.

Let us philosophize further, about their food. Know that each chaos has its two spheres, the heaven and the soil, just as we men walk on earth. Earth and heaven give us our food and the chaos is in between the two. Thus we are nourished in between the two spheres and the globules. Thus also, those who live in the water, have the earth at the bottom, and the water as chaos, and the heaven down to the water; and so they are in between heaven and earth and the water is their chaos. And their abode is according to their kind. Thus also with the gnomi, whose soil is water, and whose chaos is terra, and the heaven is their sphaera, that is, the earth stands in water. To them the earth is chaos and the water the soil. Food grows to them in such a way. The sylphs are like men, nourish themselves like men in the wilderness, on herbs in the woods. To the salamanders the soil is earth, their heaven is the air, and fire their chaos. Thus food grows to them from the earth and from fire, and the constellation from air is their heaven. Now, about the things they eat and drink, you may understand so much. Water quenches our thirst, but not that of the gnomi, nor of the nymphs, nor of the other two. Further: if water has been created for us, to quench our thirst, then another water must have been created for them, that we cannot see nor explore. Drink they must, but drink that which in their world is a drink. Eat they must similarly, according to the content of their world. One cannot find out more about these things, but only that their world has its own nature, just as ours has.

About their clothing: they are clothed and cover their genitalia, but not in the way of our world, in their own way. For they have modesty and similar qualities, as men must have, have law and similar institutions, have their authorities, like the ants, which have their king, and the wild geese, which have their leader, not according to the law of men, but according to their inborn nature. The animals have their chief, and so have they too, and more than all animals, because they are most similar to man. God has clothed all beings and endowed them with modesty, to walk and stand before man. To the beasts clothing is inborn by nature, but not to these people. To them nothing is inborn by nature; they must work for it like man whom they resemble. Their work, like man's work, is in the nature and kind of their own world and earth on which they live. For he who gave us wool from sheep, gives it to them also. For it is possible for God to create not only the sheep which are known to us, but also the same in fire, in water, in the earth. For he clothes not only us, but also the gnomi, the nymphs, the salamanders, the sylvestres. They are all under the protection of God and are all clothed and guided by him. For God has not only power to provide for man, but also for anything else, of which man knows nothing, and of which he becomes slowly aware. And when he sees and hears something, it is a miracle to him that bears no fruit, that is: he gives it no further thought, but remains obstinate and blind, like one who with good eyes has not the grace to see.

About their day, night, sleeping and waking, know that they rest, sleep and are awake like men, that is, in the same measure as man. With that, they have the sun and firmament as well as we. That is: the mountain manikins have the earth which is their chaos. To them it is only an air and no earth as we have it. From this we must conclude that they see through the earth, as we do through the air, and that the sun shines to them through the earth, as it does to us through the air, and that

they have the sun, the moon and the whole firmament in front
of their eyes, as we men. To the undinae the water is their
chaos, and water does not keep them from the sun; and just as
we have the sun through the air, they have it through the water,
in the same measure. Thus also to the vulcans through the
fire. And in the same way as the sun shines on us and ferti-
lizes the earth, know that the same happens to them as to us.
The consequence is, that they too have summer, winter, day and
night, etc. But they do not need rain, snow, etc. They have it
in a different way than we. These are the great miracles of
God. From all this we must conclude that they have pestilence,
fevers, pleurisies, and all diseases of the heaven, just as we do,
and that they follow us in all these matters, since they are men.
But in the judgment of God, in the resurrection, they will be
beasts and not men.

About their figures, know that they are different. The water
people look like men, both women and men. The sylvestres do
not conform, but are cruder, coarser, longer and stronger than
both. The mountain people are small, of about two spans. The
salamanders are long, narrow and lean. Their place and abode
are in their chaos, as was mentioned before. The nymphs live
in water, in running brooks, etc., so close that they grasp the
people who ride through or bathe therein. The mountain people
are in the mountain chaos, and there they build their houses.
This is why it often happens that one finds in the earth an
attic, vaults, and similar structures, of the height of about a
yard. They have been build by these people for their abode
and dwellings. The water people do the same in their various
places. Know also that the mountain people live in the caves
of the mountains and this is why strange structures occur and
are found in such places. These are their work. Know also about
the fire people whose yelling, hammering and working can be
heard in volcanic mountains. It can also be heard when the
elements are incinerated. For all these things are the same as
with us, but according to their secret quality. Those who travel
through wild regions learn the reason for such beings and
obtain information. There these beings are found. In the mines

also, close to good ore, etc. they are found, and in waters the same also, and the vulcans near the Aetna. And there are many more marvellous things, about their coins, payments and customs, which would be too long in this connection, but will be described in their place.

Tractatus III

HOW THEY COME TO US AND BECOME VISIBLE TO US

All that God has created, he reveals to man and lets come before him, so that man has and attains knowledge of all happenings of the creatures. Thus God has revealed the Devil to man, so that man knows about the Devil, and he has presented the spirits and other matters still more impossible for man to perceive. The angels in the heaven he has also sent down to man, so that man may really see that God has angels who serve him. Such revelations do occur, but rarely, only as often as is needed to believe and have faith in them. The same happens with the beings about whom I am writing here. They appear also. Not in order to have them all live or stay or mix with us, but God lets them wander to us and be with us as often as is needed for us to take cognizance of them and to know that God creates marvellous works. When he sends an angel to us, we learn that the Scriptures speak truthfully of them; and once we know that of one thing, we know enough forever and ever, as long as the seed of man lasts, so that it is not necessary to present things every day. And thus God has also presented these creatures to man, has allowed them to be seen, to deal with the people and talk, and so on, in order that man may know that there are such creatures in the four elements who miraculously appear before our eyes. And in order that we may have a good knowledge of things, the water people have not only been truly seen by man, but have married him and have born him children. Thus also, the mountain people have not only been seen, but people saw them, and talked to them, and received money from them and blows and similar things. The same happened with the forest people, as was mentioned before, the people saw them and dealt and walked with them. Thus also, with the vulcans, who also, as mentioned before, appeared to man, and revealed themselves essentially, who they were and what had to be understood of them all.

And so many have been held before man that from them he may draw and take an adequate philosophy, to be explored adequately in the works of God, by the light of man which has been given to man alone, beyond all other creatures. For the like must be perceived by its like, that is, man is a spirit and a man, eternal and mortal. From this reasonably follows a knowledge of the other things, namely, that he is one who has been created from God after God. Thus man can not philosophize at any time about anything unless he has a subject from which his reasoning springs and on which it is founded. If such a man is possessed by the evil spirit he must now, at any time, consider what this means. For nothing exists that remains concealed and is not revealed. Everything must come out, creature, nature, spirit, evil and good, outside and inside, and all arts, and all doctrines, teachings and what has been created. Sometimes, such things become apparent only insofar as they are preserved in the memory of the people. They are still hidden and not generally known. It is not less true that man never becomes as apparent to others as these beings become apparent to man. The nymphs, namely, become apparent to us, but not we to them, except in what they tell about us in their world, as a pilgrim would, who had been in foreign lands. They do not need such rapture, the way the mountain people enrapture us, or the water people. For they have no power over man, and their world is not such that they could adopt us. Man is not subtle in body, but coarse in body and subtle in chaos, while it is the contrary with them. Therefore, they can well stand our chaos, but not we theirs. Their element also is in itself their chaos, which to us may not be chaos. Thus they appear to us, stay with us, marry with us, die with us, and also bear children.

Now, the fact that they shall be revealed, is equal to a divine decree. God sends an angel to us, recommends his creature to him, and then takes him away again. And in the same way, these beings are presented sufficiently before our eyes. This happens with the water people. They come out of their waters to us, make themselves known, act and deal with us, go back to their water, come again—all this to allow man the contemplation of the divine works. Now, they are men, but on the animal side alone, without the soul. From this it follows that they marry men. A water woman takes a man from Adam, and keeps house for him, and gives birth to children. Of the children, we know that they follow after the father. Because the father is a man from Adam, a soul is given to the child, and it becomes like a regular man, who has an eternal soul. Furthermore, this also is well known and must be considered, that such women also receive souls by becoming married, so that they are saved before God and by God like other women. It has been experienced in many ways that they are not eternal, but when they are bound to men, they become eternal, that is, endowed with a soul like man. You must understand this in the following way: God has created them so much like man and so resembling him, that nothing could be more alike, and a wonder happened in that they had no soul. But when they enter into a union with man, then the union gives the soul. It is the same as with the union that man has with God and God with man, a union established by God, which makes it possible for us to enter the kingdom of God. If there were no such union, of what use would the soul be to us? Of none. But now there is that union with man, and therefore the soul is of use to man, who otherwise would have no purpose. This is demonstrated by them also: they have no soul, unless they enter into a union with men, and now they have the soul. They die, and nothing remains of them but the beast. And a man who is not in divine union is just like them. Just as the condition of these people is, when they are in union with man, so is the condition of man, when he is in union with God. And thus they demonstrate that they are beasts without man, and like them man without divine union is nothing. The union of two things with each other can achieve so much, that the inferior benefits from the superior and acquires its power.

From this it follows that they woo man, and that they seek him assiduously and in secret. A heathen begs for baptism and woos it in order to acquire his soul and to become alive in Christ. In the same way, they seek love with man, so as to be in union with men. With them all intelligence and wisdom are outside the qualities of the soul, and not the soul. And so they receive the soul, and their children also, by virtue of Adam's fruit, freedom and power, which holds and carries them to God. One must also think of what God will do with them in the end, since they are so close to man and must be considered as wild men; just as you say, the wolf is a wild dog, the ibex a wild he-goat and such like. One must also know that not all of them may be married to us. The water people come first; they are the closest to us. Next after them come the forest people, then the mountain and earth manikins who, however, rarely marry humans and are only obliged to serve them. The vulcans never attempt to enter into union with humans, and yet are apt to serve them. Know also that two of them, namely the earth manikins and vulcans, are considered spirits and not creatures, being looked upon as a mirage only, or as ghosts. You must know, however, that just as they appear, thus they are, flesh and blood like another man, and with that, quick and fast like a spirit, as was told in the beginning. They also know all future affairs, present affairs and the past, which are not apparent but are hidden. In that they can serve man, protect, warn, guide him, and such like. For they have reason in common with man (except in regard to the soul). They have knowledge and intelligence of the spirit (except in regard to God). Thus they are highly gifted, and they know and warn, so that man may learn about such things, and see them, and believe in such creatures. For this purpose God has let them appear to man, that he may seek knowledge and learn what God effects in such creatures.

It is said of the nymphs, that they come to us from the water, and sit on the banks of the brooks where they have their abode, where they are seen, taken also, caught, and married, as we said before. The forest people, however, are coarser than they.

They do not talk, that is, they cannot talk although they have tongues and enough of all that is needed for speech. In this point they differ from the nymphs, for the nymphs can speak in the languages of the countries; the forest people, however, cannot but they learn easily. The mountain manikins are endowed with speech like the nymphs, and the vulcans speak nothing, yet they can speak but roughly and rarely. The nymphs appear, as was said before, in human clothing, with human features and desires. The forest people are like men, but shy and fugitive. The mountain people are like men, not tall, short; sometimes they reach about half the size of man or so, sometimes more. Thus and in the same way, the vulcans appear, fiery, and fire is all over their features and clothing. They are the ones of whom one says: a fire man or spirit is going through the house; there goes a burning soul, etc. It often happens that such figures are seen. They also are the flames that are frequently observed as glowing lights in meadows and fields, running through each other and towards each other. These are the vulcans, but they are not found living with man, on account of their fire. They are often found with old women, however, that is, with witches, wooing them. One also must know that the devil takes possession of them just as he does of men, and thus goes around with them, appearing to the people in the features just mentioned. And this leads and brings them to the witches, and many such matters that happen when things are possessed by the devil, but it would be much too long to describe them here. You must know, however, that there is danger with the fire people, because they are commonly possessed, and the devil thus rages in them, which causes great harm to man. Know also that he equally takes possession of the mountain people and thus makes them subservient to him, and of the forest people also. He then can be found in the forest possessing sylphs and venturing to make love to women who live in forest regions. But they all become like lepers, scabby and mangy, and there is no help for them either.

When these beings are not possessed by the devil they are human and seek union as has been mentioned. But they keep the way of spirits, in that they disappear. One who has a nymph for a wife, should not let her get close to any water, or at least should not offend her while they are on water. And one who has a mountain manikin with him, should not offend him, particularly not at places where they get lost. But they are so much obliged to man and so closely bound to him, that they cannot get away from him, unless there is a reason for it, and this happens at the place from which they come. If one has a wife, she does not get away from him, unless she is provoked while they are on water. Otherwise she will not disappear and can be held. The mountain manikins also must keep their pledges, when they are in service and have been pledged. But obligations to them must be kept also, in all that is due to them. If duties are kept as should be, they are honest, constant, and intent on their work. And one must know that they are particularly loyal to man and much inclined to spending money. For, the mountain people have money, because they themselves coin it. This is how it must be understood. What a spirit wishes to have, he has, and their wishing and desiring is thus: when a mountain manikin wishes or desires a sum of money, if there is need for it, he has it, and it is good money. Thus they give money to many people in the galleries of mountains, to make them get away; they buy them off. All this is divine order— that they thus appear to us and that we see what is incredible to tell. Man is the most earthbound of all creatures. What he must have and wants, he must make for himself, and he obtains nothing by wishing and desiring it. But those people have what they need and desire, and man does nothing for it. They have it without work.

Tractatus IV

As we came to the end of the treatise, we had sufficiently discussed the necessities of these beings, and how they come to man. You must now, furthermore, know about their disappearance from man, and about their doings with us, with many such tales and stories that have happened with them,

17

in many queer ways. And first about those who married men and bore them children, as was said before. When they have been provoked in any way by their husbands while they are on water, they simply drop into the water, and nobody can find them any more. To the husband it is as if she were drowned, for he will never see her again. And yet, although she dropped into the water, he may not consider her dead. She is alive and he may not take another wife. If he did, he would have to pay with his life and could never return to the world, for the marriage is not dissolved but is still valid. It is the same as in the case of a wife who has run away. She is not divorced from her husband, nor he from her. It still is a valid marriage that has not been dissolved and that nobody can ever dissolve, as long as there is life. When she has dropped into the water and has abandoned husband and children, her marriage is still valid, and she will present herself on the Day of Judgment on account of her union and pledge. The soul, namely, is not taken away nor separated from her. She must follow the soul and hold to her pledge to the end. Although she remains a water woman and nymph, she must behave as the soul requires and the pledge she has given, except that she is separated from her husband, and that there is no return, unless he takes another wife. Then she returns and brings him death, as has happened many times.

It also happens that sirens are born. They are water women too, but live more on the water than in it. They are not split like fish, but are more like a virgin, yet their form is not quite that of a woman. They bear no children but are monsters, just as a strange human being can be born from two normal parents. I would like to put it thus: the water people reproduce themselves like humans, but when it happens that they produce a monster, these monsters are sirens that swim on the water, for they repudiate them and do not keep them. This is why they occur in many forms and figures, as it happens and is the case with all monsters. Thus we marvel not only at the water people but also at the sirens, who have many strange features and are very different from people. Some can sing, some can whistle with reeds, some can do this and that. Monks are also born from nymphs, that is, a monster which is shaped just like a monk. This you must know, namely, that such growths that are comparable to men and are found in some places, come from men, that is, they come from water people, earth manikins and similar beings. In order to understand the matter correctly, remember that God effects strange things in his creations. It is the same as with a comet. Born from other stars, the comet is nothing but a super-growth, that is, a growth that has not a natural course, as a star should have, but has been directed by God on purpose into another course. This is why the comet has great significance. And so have the sea wonders and similar things which thus come from water people. They too are such regular comets that God presents to man on purpose, not without meaning and significance. There is no need to write about it here, but this you must know, that great things come from such people, who should be great mirrors held before man's eyes. But love has cooled down in many, and thus they pay no attention to the signs, intent only on usury, self-interest, gambling, drinking, matters that are interpreted by these beings, as if they were saying: look at the monsters, thus you shall be after death; let yourself be warned, beware—but nothing is done about it.

To continue the discussion of these beings, know that such people also convene and assemble in one place, where they may live together and seek intercourse with man, for they love him. The reason is that flesh and blood hold to flesh and blood. There are more women than men in such groups, few men, many women; hence they are after men whenever they have a chance. From such people a group originated that is called the Venusberg. It consists of nothing else but a kind of nymphs, thrown together in a cave and hole of their world, yet not in their own chaos, in man's chaos, but in their regionibus. Know about them that they reach a very old age, but you cannot notice it, because their appearance remains the same from beginning to end, and they die unchanged. Venus was a nymph and undine who excelled others and reigned for a long time, but she

died and the succeeding Venus was not as endowed as she
had been. She died in the course of time and her kingdom
vanished. There are many tales about her. Some people believe
that she will live until the Day of Judgment, meaning: she
and her seed, not she alone. And on the Day of Judgment, all
these beings will appear before God, will dissolve and come to
an end. It is also said that those who come to them do not die
either. But this is not true, for all beings end in death and
nothing remains, neither they nor other people, nothing is without an end. It is on account of the seed that all kinds survive
to the Day of Judgment. There is a story about a different
beginning. It tells of a queen who resided there and sank into
the earth. It was a water woman who resided there. She went
into the mountain, under the pond that was above her, in her
region. There she took her abode, and for making love, she
built a gallery, for her to get to the lads and they to her.
Such strange things happened there that nobody was able to
report about these beings, what they were or where they were
from, until it came to an end. It is quite possible that this could
happen again should another nymph come, equal to her. How
often does it not happen that a man is amazingly superior to
others, and then for many years there is nobody equal to him.
Thus a special portent occurred about the nymphs, was called
the Venusberg, after the idol of unchastity. Many such strange
stories have happened on earth, but they have been greatly
despised. Yet, it should not be, because no such thing occurs
without great attention being paid to it. Hence, they should
not be despised. There is no need of writing about it here.

There is also a true story of the nymph in Staufenberg who
sat on the road in all her beauty and served the lord she had
chosen. It is quite correct that to the theologians such a being
is a devilish ghost, but certainly not to the true theologians.
What greater admonition is there in the Scriptures than to
despise nothing, to ponder everything in mature understanding
and judgment, to explore all things and dismiss nothing without previous exploration. It becomes easily apparent that they
have little understanding of these beings. Making it short, they

call them devils, although they know little enough about the devil himself. This you must know, that God lets such miracles happen not to have us all marry nymphs and live with them, but once in a while, in order to demonstrate the strange doings of divine creatures and that we may see the work of his labor. If it were the devil's work it should be despised, but it is not, for this he is not able to do; God alone is. Our nymph was a water woman. She promised herself to von Staufenberg and stayed also with him, until he married another wife, and took her for a devil. Taking her for a devil and considering her such, he married another woman and thus broke his promise to her. Therefore, at the wedding, she gave him the sign, through the ceiling, during the banquet, and three days later, he was dead. It requires great experience to judge in such matters, for the breaking of a pledge never remains without sanction, whatever it may be, to uphold honor and honesty and to prevent other evil and vice. If she had been a ghost, from where would she have taken blood and flesh? If she had been a devil, where would she have hidden the devil's marks which always go with it? If it had been a spirit, why should it have needed such a being? She was a woman and a nympha, as we have described them, a woman in honor, not in dishonor, and this is why she wanted duty and loyalty to be kept. Since they were not kept, she herself, from divine destiny, punished the adulterer (for no judge would have passed a sentence at her request, since she was not from Adam). Thus God granted her the punishment that is appropriate for adultery, and permitted her to be her own judge, since the world repudiated her as a spirit or devil. Many more such things have happened that are despised by men, badly so and it is a sign of great stupidity.

We must pay equally great attention to Melusine, for she was not what the theologians considered her, but a nympha. It is true, however, that she was possessed by the evil spirit, of which she would have freed herself if she had stayed with her husband to the end. For such is the devil that he transforms these beings into different shapes, as he also does with the witches, transforming them into cats and werewolves, dogs, etc. This happened to her also, for she never was free of witchcraft but had a part in it. A superstitious belief resulted, that on Saturdays she had to be a serpent. This was her pledge to the devil for his helping her in getting a man. Otherwise, she was a nympha, with flesh and blood, fertile and well built to have children. She came from the nymphs to the humans on earth and lived there. But then, as *superstitio* seduces and vexes all beings, she went away from her people in her superstitious belief, to the places where the seduced people come who are bewitched in *superstitionibus* and spell-bound. Mind you, she remained the same serpent to the end of her life, and God knows how long it lasted. Thus these beings are warnings to us, to make us understand what we are on earth, and in what strange ways the devil deals with us and is after us in every corner. Nothing is hidden to him, neither in the depth of the sea nor in the center of the earth, where he whiles. But wherever we are, there is also God who redeems those who are his, in all places. It is stupid, however, to consider such women ghosts and devils on the basis of such happenings and because they are not from Adam. It is holding God's works in low esteem to assume that they are rejected because they have *superstitiones*. Yet there are more *superstitiones* in the Roman Church than in all these women and witches. And so it may be a warning that if *superstitio* turns a man into a serpent, it also turns him into a devil. That is, if it happens to nymphs, it also happens to you in the Roman Church. That is, you too will be transformed into such serpents, you who now are pretty and handsome, adorned with large diadems and jewels. In the end you will be a serpent and dragon, like Melusine and others of her kind.

Therefore, consider such things carefully, and be not blind with seeing eyes and dumb with good tongues, particularly since you will not let yourselves be called dumb and blind.

Tractatus V

DE GIGANTIBUS

On Giants

You must know further that there are two more kinds that belong to the kind of nymphs and pygmies, namely, the giants and dwarfs, who are not born from Adam. Although it is a fact that Saint Christopher was a giant, yet he took birth from human seed, and hence we shall not write about him. But we shall write about the other giants, referred to in the stories of Bern, Sigenott, Hiltbrant, Dittrich, etc., also about the dwarf Laurin and others. Such stories are frequently rejected, but you must know that the same people who reject them, reject also much more important truth. They distort Christ's word and put themselves into the fore. They have much more right to disregard things that have happened in this field, but their disbelief is all the same. The giants are too strong and too strange for us. They are beyond us, and hence we are inclined to deny their existence and to consider them ghosts. The same we wish to do with Christ. He also is too strong for us, and we are inclined to do with him as with the giants, so as to have nobody who could oppose us, or nobody whom we might fear.

The beginning of theology is the enlightenment of man. He must be enlightened in all things of nature and must know them. Once he has judgment in these matters, he cannot take Christ and the Scriptures lightly but, from necessity of the inborn light, he must esteem them more highly and interpret them more broadly than the others, who see nothing but letters. This is why blind people should not touch the Scriptures, but only those who have grown up in the light of man. The others are inclined to become seducers and heretics. Therefore, learn and experience what there is to learn and experience, and do not tumble into the Scriptures, violating what seems incredible to your inexperienced brains.

And so you must have knowledge of these two kinds, giants

and dwarfs. The giants come from the forest people and the dwarfs from the earth manikins. They are *monstra* like the sirens from the nymphs. Thus these beings are born. And although it happens rarely, yet it happened so often and under such marvellous circumstances that their existence is well known and remembered. They are strange in size and strength, giants and dwarfs. There is nothing to tell about their resembling the kind of the parents from whom they were born, because they do not resemble their kind but are *monstra,* beyond the kind from which they came. Yet, they were not born in vain, but from divine order to signify something, and the significance is not small, but great. Hence, it is fair to describe their extraction, and on some other occasion to describe also their significance, why God has let them be born from people, from what causes. It is not necessary to discuss this here for it is my intention to discover the origin of all these creatures.

We shall have to tell how strange and odd they appear, with stories and deeds that they performed on earth. And also how the *monstra,* that God ordered on purpose, have died out on earth without heirs of their body and blood, so that both giants and dwarfs are gone and have died out.

First, we must write about their soul, how we can find out about it. It is thus: the parents from whom they are born have no soul, yet they are human, as has been said. From beasts that are human these beings are born. From that it follows that they have not inherited souls from their parents. They are *monstra,* in addition, and it is peculiar to their nature that they have less than their parents. From these two arguments it follows that they had no soul, in spite of the fact that they have been found performing good deeds, works, etc., sincere to each other and with qualities usually associated with the soul. But just as the parrot can talk and the monkey imitate man and many such things occur, their innate nature is equally able to perform and accomplish such things. That they should have a soul, however, cannot be concluded from these facts. God who can endow with a soul whomever he wishes, may also give these people a soul as he did to others. We mentioned the union between man and God and that between nymphs and men. And yet, they have not been born for the sake of the soul, but for the sake of the creature that God may be recognized the more marvellously in his creations; not that God wanted souls that were giants, since one man is like another in the kingdom of God. This is why I still consider them beasts whose soul is unknown to me. And although they have also performed good deeds, yet I cannot feel that they were seeking salvation; this would be hard to believe, but rather that they acted like clever animals. Truly, if a fox or a wolf could speak, they would not be very different. One must concede a great deal to natural understanding that does not serve the purpose served by one who has a soul.

The forest people, that is, the air people, make the giants when they come together in intercourse, just as a conjunction produces a comet, an earthquake or similar happenings. When such a monster shall be born, it is not born following the course of nature, but against the rules, by special divine providence. There is no need to write about this here. Thus, these forest people join together and bear monsters, but we must write how and in what way. It cannot be understood except as a divine work and order that can be explained only in analogy with astronomy, when we remember the comets and the production of things that are not in the regular course of nature, yet are produced, such as the earthquake and similar happenings. They come and then do not recur for a long time. Thus we must also understand the giants. They are produced in the same way, by a constellation that is in men, not in heaven, for the heaven has nothing to do with the production of monsters such as these. The people are no concern to the heaven in its constellation, although they all live under the heaven and are confined by it, but they have a *corpus* and another chaos. Therefore the heaven cannot impress anything on them, for it does not stick. But if you can understand the natural course of the heaven, you can understand also an independent one, its rise, death and decrease. Everyone must know that if a particularly tall or very small person is born among us, this must not be attributed to the heaven but to nature's own course. There is no need to write about this here.

About the dwarfs you must know that they are born from the earth manikins in the mountains. Hence, they are not as tall as the giants but are smaller in the same proportion as the earth manikins are smaller than the forest people. They are monsters also like the giants and their birth must be understood the same way. Thus we can highly appreciate the words of Saint John the Baptist who said to the Jews: And think not to say within yourselves, because you are the children of Abraham, you can resist God and defy him, as if you were not his own and he were nothing, as if he could not make other men. For I say unto you, that God is able of these stones to raise up men. This must also be understood to mean that although we are from Adam yet there are people who are not from Adam, such as giants and dwarfs who are greater and stronger than we. It also means that if you shall not do honest work, God can exterminate you in the root and let you all dry out like fruit on a tree, and create other people thereafter. If it was possible for him to make Adam and his children from limus, it is also possible for him to make without limus other people such as nymphs, giants, etc., and to populate and maintain his world with them. Thus they stand as a warning that we are not alone and cannot force God. If he can do one thing, he can do another also. If he can make a man seven feet long, he can also make him twenty or thirty feet long. This is apparent with the giants and similar beings, that they all are warnings to make us realize that God is the Lord, who can do things and endow a being with a soul in one breath.

To come to an end, know that such people, giants and dwarfs, may well make women from Adam pregnant, because same likes to meet same, and nature may permit it. Yet you must know that the monsters are not fertile, although they are monsters only so far as their strength and size is concerned; they are not misbuilt. But they have only one seed available, that is, they do not attain the third, fourth, etc. generation. Know also that when they get into man's kind, the result is uncertain.

If the child's body takes after the mother it falls into her kind; if not, it becomes a beast like the father. It is not possible for it to become a mixed being because the seed of one of the two always prevails. It happens otherwise that both become one seed, but here not. And if it were one seed it would have to be qualified by the one partner who gives the soul. More is said about this subject in other philosophical treatises, and there is no need to tell it here. Their kind, however, has died out and they have left no descendants to replace them. The comets, likewise, leave no seed behind them but come straight on and off at times, and at times up from below, all of a sudden, and then disappear.

Tractatus VI

ON THE CAUSES OF SUCH CREATURES

Now that we must reflect in terms of natural philosophy about the causes of these creatures we must remember the facts, all the points, many of which have been discussed in the previous treatises. We need not repeat them here. Further causes that may be explored are these: namely, that God has set guardians over nature, for all things, and he left nothing unguarded. Thus gnomes, pygmies and *mani* guard the treasures of the earth, the metals and similar treasures. Where they are, there are tremendous treasures, in tremendous quantities. They are guarded by such people, are kept hidden and secret so that they may not be found until the time for it has come. When they are found, people say: in times of old there used to be mountain manikins, earth people here, but now they are gone. This means that now the time has come for the treasures to be revealed. The treasures of the earth are distributed in such a way that the metals, silver, gold, iron, etc., have been found from the beginning of the world on, and are being discovered by and by. They are guarded and protected by these people so that they may not be found all at once, but one after another, by and by, now in this country, now in another. Thus the mines are shifted in the course of time from country to country and are distributed from the first day to the last. The same applies also to the fire people. They too are guardians, of the fireplaces, in which they live. In these places the treasures, that others guard, are forged, prepared, made ready. When the fire is extinguished, it is the earth manikins' turn to be on guard. And after the earth manikins' guard, the treasures are revealed. It is the same with the air people. They guard the rocks that lie on the surface, that have been made by the fire people, and put at the place where they belong, from where they get into man's hands. They guard them so long, until the time has come. Wherever there are treasures, such people are on hand. These are hidden treasures that must not be revealed yet. Since they are guardians of such matters, we can well understand that such guardians are not endowed with a soul, yet similar to and like men. The nymphs in the water are guardians of the great water treasures, that lie in the sea and other waters, that have also been melted and forged by the fire people. It is, therefore, commonly understood that where nymphs are, there are considerable treasures and minerals and similar matters which they guard. This is apparent in many ways.

The cause of the sirens, giants, dwarfs, also of the will-o'-the-wisps, who are monsters of the fire people, is that they predict and indicate something new. They are not on guard, but signify that misfortune is threatening people. Thus when lights are seen it means the impending downfall of that country, that is, it commonly signifies the destruction of the monarchy and similar things. Thus the giants also signify great impending destruction of that country and land or some other such great disaster. The dwarfs signify great poverty among the people, in many parts. The sirens signify the downfall of princes and lords, the rise of sects or factions. For God wants us all to be of one essence. What is against it, he drops. And when this is going to happen, signs occur. These beings are such signs, as has been said, but not they alone; there are many more. You must know that the signs change each time. They do not appear in one way, but are hidden to our eyes.

And, finally, the last cause is unknown to us. But when the

end of the world will come close, then all things will be revealed, from the smallest to the largest, from the first to the last, what everything has been and is, why it stood there and left, from what causes, and what its meaning was. And everything that is in the world will be disclosed and come to light. Then the fake scholars will be exposed, those who are highly learned in name only but know nothing by experience. Then, the thorough scholars and those who are mere talkers will be recognized for what they are, those who wrote truthfully and those who traded in lies, the thorough and the shallow ones. And to each will be measured according to his diligence, earnest endeavor and truth. At that place, not everyone will be or remain a master, or even a doctor, because there the tares will be separated from the wheat and the straw from the grain. He who now cries, will be quieted, and he who now counts the pages, will have his quills taken away. And all things will be revealed before the Day of Judgment breaks, order that it be found of all scholars, from the past to that very day, who had knowledge and who not, whose writings were right and whose wrong. Now, in my time, this is still unknown. Blessed will be the people, in those days, whose intelligence will be revealed, for what they produced will be revealed to all the people as if it were written on their foreheads. For that time I also recommend my writings for judgment, asking that nothing be withheld. Thus it will be, for God makes the light manifest, that is, everyone will see how it has shone.

LITERATURE

1. BIBLIOGRAPHIES

The manuscripts and editions of the works of Paracelsus and the chief literature about him from 1526 to 1893 are listed in Karl Sudhoff's monumental bibliography:

Karl Sudhoff, *Versuch einer Kritik der Echtheit der Paracelsischen Schriften.* I. Theil. Die unter Hohenheim's Namen erschienenen Druckschriften. Berlin, 1894: Bibliographica Paracelsica, Besprechung der unter Theophrast von Hohenheim's Namen 1527-1893 erschienenen Druckschriften; II. Theil. Paracelsische Handschriften I. Hälfte. Berlin, 1898: Paracelsus-Handschriften gesammelt und besprochen; II. Hälfte. Berlin, 1899.

The literature from 1893 to 1932 with additions to the earlier periods was compiled by Karl Sudhoff in:

Acta Paracelsica. Im Auftrag der Paracelsus-Gesellschaft, herausgegeben von Ernst Darmstaedter, Richard Koch, Manfred Schroeter. Heft 1-5. München 1930-32, Verlag der Paracelsus-Gesellschaft.

In the United States there are two important Paracelsus collections of which checklists have been published:

The Paracelsus Collection of the St. Louis Medical Society [Robert E. Schlueter Collection], *Bulletin of the History of Medicine,* 1941, Vol. 9, pp. 545-579.

Paracelsus Library of Dr. Constantine Hering, *The Hahnemann Medical College of Philadelphia,* 1932, pp. 7-18.

2. WORKS OF PARACELSUS

a) In original German and Latin

Johannes Huser, *Die Bücher und Schrifften des . . . Paracelsi.* 10 Volumes, Basel, Conrad Waldkirch, 1589-91.

Karl Sudhoff and Wilhelm Matthiessen, *Theophrast von Hohenheim gen. Paracelsus Sämtliche Werke*: 1. Abteilung, Medizinische naturwissenschaftliche und philosophische Schriften. 14 Volumes. Munich and Berlin, 1922-1933. II. Abteilung, Die theologischen und religions-philosophischen Schriften. 1 Volume. Munich, 1923.

b) Translation into modern German

Bernhard Aschner, *Paracelsus Sämtliche Werke,* nach der 10 bändigen Huserschen Gesamtausgabe (1589-1591) zum erstenmal in neuzeitliches Deutsch übersetzt. Mit Einleitung, Biographie, Literaturangaben und erklärenden Anmerkungen versehen. 4 Volumes. Jena, Verlag von Gustav Fischer, 1926, 1928, 1930, 1932.

c) English translations

In the 16th and 17th centuries a number of alchemical and therapeutic treatises were translated into English. They are listed in Sudhoff's Bibliography and in Bayard Q. Morgan, *A Critical Bibliography of German Literature in English Translation, 1481 - 1927. Supplement 1928 - 1935*. Second edition. Stanford University, California. London, Oxford University Press, 1938.

> See also: Arthur Eduard Waite, *The Hermetic and Alchemical Writings of Aureolus Philippus Theophrastus Bombast, of Hohenheim, called Paracelsus the Great*. Vol. I. Hermetic Chemistry. Vol. II. Hermetic Medicine and Hermetic Philosophy. London, James Elliot and Co., 1894.

3. LITERATURE ON PARACELSUS

a) In German

A short, but the most authoritative, biography is: Karl Sudhoff, *Paracelsus, Ein deutsches Lebensbild aus den Tagen der Renaissance*, Leipzig, 1936. It contains most of the documents to his life.

The iconography of Paracelsus is described in: Carl Aberle, *Grabdenkmal, Schädel und Abbildungen des Theophrastus Paracelsus*. Salzburg, 1891.

> Raymund Netzhammer, *Theophrastus Paracelsus, Das Wissenswerteste über dessen Leben, Lehre und Schriften*. Einsiedeln, 1901.
>
> R. Julius Hartman, *Theophrast von Hohenheim*. Stuttgart und Berlin, 1904.
>
> Joh. Daniel Achelis, *Paracelsus Volumen Paramirum (Von Krankheit und gesundem Leben)*. Jena, 1928.
>
> Friedrich Gundolf, *Paracelsus*. Berlin, 1928.
>
> B. S. von Waltershausen, *Paracelsus am Eingang der deutschen Bildungsgeschichte*. Leipzig, 1936.
>
> Franz Strunz, *Theophrastus Paracelsus, Idee und Problem seiner Weltanschauung*. Salzburg and Leipzig, 1937.

b) In English

> Franz Hartmann, *The Life of Philippus Theophrastus Bombast of Hohenheim, Known by the Name of Paracelsus*. London, 1887.
>
> Anna M. Stoddart, *The Life of Paracelsus, Theophrastus von Hohenheim, 1493-1541*. London, 1911.
>
> John Maxson Stillman, *Theophrastus Bombastus von Hohenheim Called Paracelsus, His Personality and Influence As a Physician, Chemist and Reformer*. Chicago and London, 1920.

Milton Keynes UK
Ingram Content Group UK Ltd.
UKHW040702110824
446772UK00001B/94